CRYSTALS

CRYSTALS

Ken & Joules Taylor

C&B

COLLINS & BROWN

Dedication

For the Earth that provides the minerals from which our bodies are made
For those nameless supernovae from which the Earth was made
And for you – the reader – who, no less than the planet and the stars,
Provides the potential from which the future is made

First published in Great Britain in 1999
by Collins & Brown Limited
London House
Great Eastern Wharf
Parkgate Rd
London SW11 4NQ

9 8 7 6 5 4 3 2 1

British Cataloguing-in-Publication Data: A catalogue record for this book is available from the British Library.

ISBN: 1 85585 702 2 (Boxed set edition)
ISBN: 1 85585 686 7 (Paperback edition)

Conceived, edited and designed by Collins & Brown Limited

Distributed in the United States and Canada by Sterling Publishing Co, 387 Park Avenue South, New York, NY 10016, USA

Editor: Amanda Lebentz
Designer: Roger Daniels
Artwork: Lucy Bristow

Reproduction by Media Print Services, London and Colour Symphony PTE Ltd, Singapore
Produced through Phoenix Offset
Printed and bound in China

IMPORTANT
The views expressed throughout this book are not intended to reflect those of the medical establishment. The authors and publisher can take no responsibility for any accidents or ill-health sustained while using crystals in any of the situations described or pictured in this book. Also please bear in mind that the suggestions in this book are in no way intended as a substitute for advice or medical treatment from your doctor or any other health specialist. It is also important to keep any small samples of crystals which can be easily swallowed, as well as any crystals with sharp edges, away from young children.

Contents

The magic of crystals

The beauty and power of crystals and other stones have long been recognized, from the ancient times of Tutankhamen to the modern masterpieces of Cartier and Fabergé. From precious jewellery to the bricks and mortar that make up our homes, they have now become an integral part of our lives.

This book is a guide to your personal discovery of crystals. Divided into four sections – Crystals for the Home, Crystals for Health, Star Crystals and Crystals for Personal Power – there are hundreds of project ideas and inspirational tips for you to dip into. You can use the magic of crystals to enhance and re-balance every aspect of your life.

What are crystals?

The word 'crystal' derives from the Greek for 'ice', and refers to the early belief that this stone was simply water which had frozen so solid that it was beyond thawing. As late as the 18th century some scientists maintained that clear rock crystal was simply fossilized ice!

Banded fluorite

Of course, scientific knowledge then was not as advanced as it is now. These days, we know that one atom of any mineral is the replica of another atom of the same type. Gold atoms, for example, are all identical – and this is the basis of mineral classification.

A mineral that is composed solely of one type of atom is called an element, but most minerals are composed of a variety of different sorts of atoms, glued together by electrical forces to produce molecules. A molecule of water, for example, is composed of the atoms of two elements: hydrogen and oxygen. The crystal form of water is ice, a solid substance in which the atoms of hydrogen and oxygen are arranged in ordered geometrical patterns.

We see the natural shape of water crystals in the snowflakes that drift from the sky in the freezing winter. They grow, floating on the breeze, by collecting more and more molecules of water, which arrange themselves according to the characteristic hexagonal pattern of their atomic structure.

Water is a mineral like any other. In polar regions, the hills, mountains, cliffs and even the 'soil' itself consist of solid, crystalline water. Ordinary rocks are no different except that the temperature at which they thaw is so high that we only see molten rock in the natural world in an active volcano.

Some minerals have the cube as their basic building blocks, some have the pyramid formation, others are patterned on the hexagon; the wide variety of forms and shapes is one of the things that makes crystallography such a fascinating subject.

Because of the high temperatures involved, many crystals are formed deep in the earth or in volcanoes, where mineral-rich liquids slowly solidify in fractures in the earth's crust. It is because different minerals freeze at different temperatures that we find such spectacular crystals as tourmalinated quartz, where tourmaline crystals seem to have penetrated the rock crystal like magical arrows. Of course, it is the other way around –

'Ice-like' colourless crystals of quartz, and brownish dolomite crystals, found in Durham, England.

the tourmaline crystallized while the quartz was still molten and, when it too crystallized at last, the quartz simply enveloped the tourmaline. Not all crystals rely on volcanic heat to form. Some minerals are soluble (salt, for example, dissolves readily in water) and when the mineral concentration in the liquid is high enough, the atoms start to stick together, forming crystals. Selenite, which can form distinctive desert roses (see below right), is the first of various minerals, including salt, to crystallize when sea water evaporates.

To further complicate the story of how crystals grow, it is worth noting that many minerals can crystallize in a variety of different shapes and forms. At first glance the graphite 'lead' in a pencil bears no relation to the glittering diamond in a ring, but they are actually the same atomic element – carbon. What has happened is that the diamond has been subjected to enormous pressure underground and its atomic structure has been crushed into a more compact crystal lattice.

A NEW INTEREST IN STONES

Perhaps the huge upsurge of spiritual interest in crystals towards the end of the 20th century coincides with the significance many people attach to the dawn of a new millennium. Or perhaps, more people have become intrigued by the new and interesting crystals now available on the market from many corners of the world. Whatever the reason, there is certainly no doubting the fact that precious stones have been a lucrative trading commodity for well over 5,000 years.

Apart from useful stones such as flint, which was mined and extensively trafficked in the Neolithic Age, even mere curiosities like amber were transported for thousands of miles in the Bronze Age, on trade routes that reached from the Baltic to the Mediterranean. The beauty and uniqueness of crystals has long been appreciated and this value is still high.

But there is another facet to crystals. The search for new sources of power and energy has always been a driving force, particularly in the modern world. The 19th century was shaped by the exploitation of coal as a means of providing steam power, and today the development of crystal-based laser technology used, for example, in modern medicine for eye surgery, has been a crucial breakthrough for our society. It is not surprising then, that we are more and more interested in the power of crystals as a source of empowerment in our personal lives. These stones are the product of natural energy derived from natural forces such as ancient stars, the sun, the movement of the earth over thousands of years and the heat of molten rock. It would almost seem wasteful to ignore them – something that we are becoming less willing to do in this age of conserving energy.

Desert rose

Crystal forms

Crystals are formed by a cornucopia of ingredients, all influenced by specific conditions of temperature, pressure, space and time. Here is a selection of some of the most popular forms and types of crystals.

A **geode** forms when a mineral solution fills a rock cavity and crystallizes, then the enclosing rock weathers away. Geodes come in many shapes and colours, depending on the host rock and the minerals present.

A **phantom** is formed when a crystal undergoes two (or more) distinctive phases of growth, such as when there is a change to the mineral content of the solution that is crystallizing, giving the growing crystal a different colour.

Natural **fluorite** crystals grow as a series of interlocking cubes. They have the same octahedral pattern as diamonds, producing a distinctive double pyramid shape.

The hexagonal cross section of **aragonite** is composed of a series of smaller crystals that have grown side by side, their touching edges 'fusing' together.

This **desert rose** grew when gypsum crystallized out of mineral-saturated water during the evaporation of ancient lakes: many specimens are powdered with fine sand.

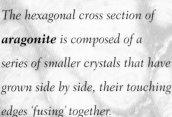

Double terminator crystals have a naturally faceted point at each end of the shaft, sometimes formed when a crystal shaft snaps and subsequently grows a second pointed end.

Stilbite crystals are found in veins and cavities in basalt lava, their characteristic shape – like a bound sheaf of wheat – is formed by an aggregate of crystals.

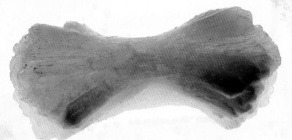

Celestite usually occurs when mineral deposits are formed by the evaporation of water. However this crystal also grows in clusters lining cavities in sedimentary rocks.

The **rock crystal wand** is a form of quartz, the most abundant mineral in the earth's crust. Huge crystals measuring several metres in girth have been found, which are thought to have taken 250,000 years to grow. Brazil is a noted exporter of rock crystal.

Pyrites occur in a variety of forms ranging from perfect cubes to spheres and nodules, the latter growing as myriad needle-like crystals radiating from a central point.

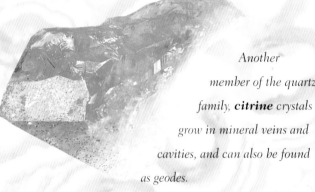

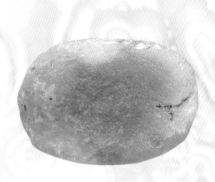

Another member of the quartz family, **citrine** *crystals grow in mineral veins and cavities, and can also be found as geodes.*

Opal *is a solidified gel containing up to 10 per cent water, which has no crystal structure. Australia is a notable source of opal, particularly black opal.*

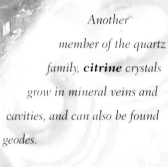

Rutilated quartz *is rock crystal that contains inclusions of needle-like crystals of rutile (titanium oxide).*

Like other members of the chalcedony family, **blue lace agate** *is composed of minute quartz crystals. The characteristic swirls reveal different phases of growth.*

Amethyst's *purple colour derives from minute quantities of iron, which has been incorporated into the crystal. It is most commonly found in volcanic rocks.*

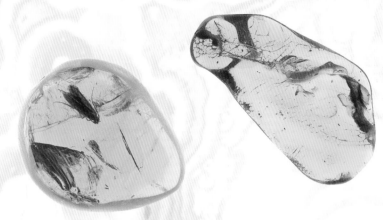

*Although originally associated with Labrador, **spectrolite** is also found elsewhere, notably in Finland. Its crystal structure has layers of weakness between which tiny spaces may develop, giving rise to the characteristic play of light.*

*While many specimens of **amber** are crystal clear, this fossilized resin from ancient pine forests often envelops and preserves scraps of foliage or even insects. For millennia, the Baltic has been a source of high-quality amber.*

*The colours in **banded fluorite** are created over a period of time during which the crystal's new growth is dyed by tiny amounts of a succession of different minerals.*

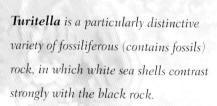

*Turitella** is a particularly distinctive variety of fossiliferous (contains fossils) rock, in which white sea shells contrast strongly with the black rock.*

*Heating amethyst turns its purple colour to yellow. **Ametrine** is artificially produced by carefully heating amethyst until some, but not all, of the colour is transformed.*

Crystals, colour and light

The colour of many transparent crystals is determined by minute traces of other minerals that suffuse the growing crystal. For example, ruby and sapphire are both made of corundum, which is colourless in its pure state. The reason they look so different is because ruby is coloured by chromium, whereas the other colours of corundum (which include yellow, green, pink and purple, as well as blue), are produced by the addition of iron and titanium in varying proportions. Many minerals, however, are opaque – which means that light cannot shine into or through them (hematite, turquoise, and basalt are examples). Others are translucent, allowing only some light to shine through them (moonstone, opal, and tiger's eye belong to this category). Some minerals produce such a range of crystal forms that specimens fit every category. Quartz, for example, is transparent in its pure form of rock crystal, translucent as carnelian and opaque as jasper.

WHAT IS REFRACTION?

When light passes into a transparent mineral it is bent, or refracted, a feature demonstrated by plunging a pencil halfway into a glass of water so that the shaft appears to bend at the point where it enters the liquid. Some minerals bend light more than others, for example, diamond, which not only has a high degree of refraction but also a powerful ability to spread different wavelengths of light. Longer wavelengths (those toward the red end of the spectrum) pass through minerals with relatively little refraction, while shorter wavelengths (the blue/violet end of the spectrum) bend further. This effect (known as dispersion) accounts for the splitting of light into the spectrum by a prism, the appearance of the rainbow, where light is refracted through the drops of rain, and the fire that sparkles from diamonds. Many

This cut diamond illustrates exceptional light dispersion, producing a fiery effect.

minerals (those whose crystal structures are not cubic) possess double refraction, whereby a beam of light entering them is split in two. Optical calcite is a good example. Double refraction also causes dichroism, an optical property that causes a crystal to have one colour (or shade of colour) when looked at in one direction, and a different colour when seen from another angle. These minerals absorb light differently according to what part of the crystal lattice the light is travelling through. Kunzite and iolite both display strong dichroism.

Many other fascinating optical effects are exhibited by nature's wide variety of crystals. The iridescence of spectrolite and the glow of moonstone, for example, are attributed to layers of tiny inclusions that act like banks of mirrors, reflecting the refracted light. Minute flakes of the mineral mica also act like mirrors, and give stones such as aventurine their sparkle. Some minerals (such as certain specimens of rose quartz) have microscopic canals in their crystal structure, which are arranged at precise angles to one another, so that they channel light to produce distinctive rayed star patterns.

Light rays can also interfere with each other when reflected by an ultra-thin film, producing beautiful 'nebulous rainbows'. This light show (like oil on water) is perhaps best seen in opals, but fine cracks in rock crystal can show an enchanting display – and on a much larger scale.

TERMINOLOGY

• *Asterism* Shining lines that cross one another like rays from a star, caused by light reflected by a series of microscopic canals.

• *Chatoyancy* Bands of light and dark that vary in width and hue as the stone is turned in the light.

• *Cleavage* A mineral's tendency to split when force is applied, leaving a flat surface. This happens when there are layers of weakness within the crystal.

• *Cryptocrystalline* Composed not of a single crystal but of a myriad of minute crystals, fused together.

• *Dichroism* The way different colours or shades can be seen according to the angle at which you look at the crystal.

• *Double refraction* Light is split into two rays, producing a double image.

• *Double terminator* A crystal which has naturally faceted points at both ends.

• *Pyroelectricity* Static electricity produced when a stone is heated or rubbed, e.g. it attracts dust.

• *Facet* Cut into the stone at an exact angle, a facet is expertly created to show a stone's particular refractive properties to best effect.

• *Inclusion* Any embedded foreign body, such as tourmaline crystals in quartz, or a moth in amber.

• *Labradorescence* The colourful play of refracted light reflected from cleavage planes.

• *Optical lens* Refracted light is bent so as to enlarge, reduce, invert, or otherwise distort an image.

• *Piezoelectricity* The static electricity that is produced when the crystal structure is deformed, for example by banging.

• *Schiller* A glowing sheen caused by internal reflections, which moves as the stone is turned.

• *Varicoloured* Crystals grown in the presence of varying impurities (imparting different hues).

• *Window quartz* Extra natural facets, usually diamond or lozenge-shaped, that occur where the sides of the crystal meet the facets that taper towards the point.

Buying crystals

The blossoming interest in crystals and other minerals means that there are many more outlets selling crystals, from little curiosity shops to market stalls to mail-order companies. The prices you pay for stones depends upon their rarity, size and form. A small natural specimen of amethyst, for example, will cost far less than a large hand-carved amber figurine.

Most people build up their collections over a period of time, purchasing a new crystal as and when they can afford it and slowly gaining experience of what represents good value for money. Of course, you may be given gems as presents or might enjoy prospecting for minerals, which can be an absorbing hobby in its own right. Some minerals can be collected quite easily: for example, many can be picked up from beaches.

Although you are unlikely to come across fake gems when making a purchase, it is worth bearing in mind that over the centuries, sharp-practising miners and merchants have exploited techniques to transform minerals into something more 'desirable' – or more expensive!

A favourite trick is simply to dye the specimens a darker or even a different colour. Turquoise and chalcedony are particularly vulnerable to dye. Then there are various forms of heat treatment that can be applied: citrine, for example, may be produced from either smoky quartz or amethyst. Sometimes, these alterations are only temporary and the

It is quite permissible to buy a stone on impulse. If you spot a stone that is truly irresistible, then it has already worked a little magic

stones can degenerate horribly within a very short space of time.

While acquiring a talismanic crystal should be a little more special than buying groceries, it is quite permissible to buy a stone on impulse. If you spot a stone that is truly irresistible, then it has already worked a little magic. The fact that it is capable of evoking a strong, spontaneous attraction means that it has struck a chord deep within you.

The emotion it conjures up – known as its resonance – is one of the most important keys to using crystals as talismans. Treat the experience with respect: enjoy it, but don't abuse it. For example, if you find a stone to help overcome shyness, be wary of the temptation to shout it from the roof-tops. If it's a stone to help you learn, don't go out celebrating instead of revising for that important exam, or you'll certainly learn something – the hard way!

BEFORE YOU BUY

When you have identified your talismanic project (see pages 138–141), you can tune into the resonance of a stone even before you set out to buy it.

To do this, you need to break your normal routine in order to begin building a bond before arriving at the shop. For example, if the stone you desire is one with invigorating properties, then you should take some

When choosing a crystal, pass the palm of your hand over a selection and try to ascertain which stone gives off a feeling of warmth.

energetic physical or mental exercise, or go somewhere that's buzzing with activity. If you are planning to buy a stone with calming properties, then it's a good idea to visit peaceful places such as a park, library or gallery, or just take some time out and rest. Getting in the mood before you take your pick from the stones on offer can help prevent potentially costly mistakes.

Often one specimen may stand out in a tray of similar stones – and you will know at a glance that it's the right one for you. At other times, you may need to pass the palm of your hand over the selection, looking for a feeling of warmth emanating from the stone with the most harmonious resonance. Alternatively, you can handle each stone individually,

*O**ften one specimen may stand out in a tray of similar stones – and you will know at a glance that it's the right one for you***

feeling for a subtle tingling in your fingertips: the best stone for you is the one that produces the strongest sensation. Some people mentally 'ask' the stone how it feels about going home with them, using extra-sensory perception to assess its response. Occasionally, you might simply need to buy the one which is least easy to leave behind.

It is best to avoid buying stones with conflicting natures at the same time. If this is unavoidable, then at least carry them in different containers or pockets – and be aware that you may be prone to emotional turbulence or unaccustomed mood swings on the way home!

Bringing your crystals home

When you arrive home with your new stone, pause to savour the occasion before you unwrap it. It is as if you are about to introduce a new friend and invite them to make themselves at home with you.

You don't actually have to straighten the cushions and make dinner – but you could certainly set the mood. Try playing some music suited to the nature of the stone. For example, if the stone inspires sensitivity, put on something quiet and soothing, or if it excites the passions, choose a rousing, upbeat tune.

If appropriate, you should open the package in the room where the stone is intended to have its greatest effect. Just as you are about to unveil it, consider for a moment the long journey that has brought the gem to its present location. Think about its origin, perhaps geological aeons ago in a country or sea that no longer exists, and imagine how and where it travelled up until the time that you first set eyes on it.

As soon as you open the package, engage either in the activity you hope it will promote or in a dramatic reconstruction (in your imagination if needs be) of that activity. For example, you might bring home a rock crystal wand because you aim to learn more about crystals so that you can work with them – whether in the home, for healing, or for personal empowerment.

Gather together a selection of books and study materials, but before you embark on a study period, remember: learning should be fun. So enjoy the experience, and set the mood with quiet music and scented incense. Keep a pen and paper handy to make a note of any ideas that might crop up about your way of studying, for example, the best times and places for you to study in the future. Even the briefest attempt will empower the stone and help to make it more effective later.

Stones can quickly become charged with talismanic energy, and having them with you when you engage in the activity for which you bought them will strengthen their power.

Especially during the early stages of building a collection, it is well worth spending a few minutes examining every new addition in the company of each of your existing stones. Not only does this perform a sort of social introduction – and there is no need to feel embarrassed about entertaining this idea – but it can also help you to discover or reacquaint yourself with any overlooked benefits (physical and metaphysical) of the specimens you already possess.

CARING FOR YOUR CRYSTALS

Nothing lasts forever – not even diamonds, which are all gradually turning to graphite – so caring for your collection is obviously the key to keeping them in good condition for as long as possible.

Some minerals, such as pyrites, can begin to degrade as soon as they are taken out of the ground; they simply rust when exposed to air and the moisture it contains. Others, such as rose quartz, can start to lose their colour if exposed to heat or strong light. Opals will lose their famous irridescence if handled too much (the grease from skin can actually penetrate and eventually dull this porous stone).

Some minerals, such as halite (salt) will start to dissolve if washed in water. Most, however, are very robust and, like the mountains from which they come, will easily outlast us!

It is rare for minerals to need more than a little light dusting from time to time and, in general, the less fuss you make the longer they last. It is purely a matter of personal preference whether to keep them all on permanent show (in an open bowl for example), or to treasure them carefully within a jewellery box.

AVOIDING MISHAPS

One point to bear in mind is that some minerals are softer and more fragile than others. It is only common sense when transporting them to prevent them from rubbing or banging against one another (scratches and chips can occur in seconds but the damage and regret can last a lifetime). It is always prudent to check at the time of purchase if any of the stones will require any special care and attention.

If you plan to use them for healing (especially if you are experimenting with crystal essence remedies, where they are steeped in water and the water is imbibed, see pages 70–73), it is particularly important to get advice from the vendor about any problems in using the specimens you are thinking of buying.

A jewellery box is ideal for storing your favourite or most precious stones – especially those that are soft or fragile.

One of the greatest joys is sharing your interest with friends, so carry a number of your favourite specimens around with you, and leave some on display in your home.

Crystals make excellent conversation points and, as their popularity continues to expand into the 21st century, the opportunities for swapping stones, exchanging information, making new contacts and generally enjoying stimulating company will also continue to grow.

CRYSTALS FOR THE HOME

Small or large, old or modern, terraced or detached, any home can be transformed by crystals. By choosing stones for certain situations, you can enhance different aspects of a room to create a comfortable environment. This chapter tells you all you need to know about displaying crystals to make the most of your home, from using rock crystal clusters to brighten a dark hallway to warming up the living room with amber. Discover how crystals work with feng shui to promote harmony and take inspiration from a host of decorative ideas. There are myriad ways to introduce crystal power into your home, and you don't have to be an expert in order to reap the benefits.

Foundations of stone

Over the millennia, mankind has called many types of dwelling home, from caves to condominiums. Ancient civilized societies raised substantial edifices of stone, the incredible ruins of which can be seen today. There is the red city of Petra in Jordan, hewn from living rock; Machu Picchu, the lost Andean city of the Incas in Peru; and Angkor Wat, the city of 100 temples and ageless mystery in Cambodia. These fascinating reminders of ancient civilizations and architectures cannot fail to inspire awe. Then there is Stonehenge, the prehistoric marvel of celestial architecture and, of course, the classical glories of Egypt, Greece and Rome. The sight of these stone structures can evoke unfamiliar and powerful sensations.

The materials from which our own homes are made can also affect us: most people can discern a difference in atmosphere between a cottage built of local stone and a tower block of steel and reinforced concrete.

From the structure of our walls to the glass in our windows; from the metal nails and screws that hold everything together to the wires and plumbing that run like nerves and arteries under a skin of plaster and paint, a house is a construction of rock and metal. In fact, it is just like jewellery. The rocks differ, of course – jewellery being made up of gemstones and precious metals instead of base ones – but the parallels are worth considering, and even the prices can be comparable.

While houses are the shells we build for protection from the weather and as a refuge from the outside world, homes are more intimate, affording us the luxury of moulding our personal environment to suit our individual tastes. Home is a place where we can truly be ourselves. Yet most people don't have the option of owning the home of their dreams or the resources to splash out on redecoration. Many people live in rented accommodation and may not have permission to decorate as they please. One of the wonderful bonuses of using stones is that you don't have to spend a fortune on expensive schemes or furnishings to dramatically improve your home.

Tiny, multi-coloured stones make an attractive gravel-like covering for house plants, helping the compost to retain moisture and keep the plants healthy.

KEY USES FOR CRYSTALS AND GEMS IN THE HOME

PROTECTION

The home provides a measure of security for those within its walls and protection from the weather, enemies or strangers. Hematite is unsurpassed in its qualities as a protective gem.

Hematite

WARMTH

Warmth ensures a pleasant living environment, allowing for comfort, relaxation and recuperation from the outside world. Imperial topaz represents the life-giving hearth.

Imperial topaz

NATURAL LIGHT

It is easy to take light for granted but an inadequate supply can have a profound effect upon us, not merely psychologically but physically. Citrine captures the essence of sparkling sunlight.

Citrine

FRESH AIR

Air should flow freely through the home, reviving everything it touches. The clarity of rock crystal epitomizes the element of air.

Double-terminator rock crystal

CLEANLINESS

Hygiene is important in the home to protect health and wellbeing. The antiseptic properties of blue lace agate are ideal in an area where cleanliness is a priority, such as in the kitchen.

Blue lace agate

COLOUR

The way in which we decorate the home reflects our personality, and being surrounded by our favourite colours can help sustain emotional and mental health. Opal, in its various coloured forms, may help to inspire decorating ideas.

Opal

VIEWS

The direction in which your home faces has a bearing both on how it is decorated and what rooms are used for. The view(s) may also be a significant factor in the wellbeing of the inhabitants. Moss agate may inspire solutions.

Moss agate

SPACE

Whatever the size of the home or garden, making the most of the available space is a priority. A single aqua aura crystal, hung where it best catches the light, may prove particularly enlightening.

Aqua aura

NEIGHBOURS

Often a prickly subject! White moonstone placed against an intervening wall will help to ease friction. In confrontational situations, holding a tumble-polished rock crystal may also prove useful.

Moonstone

The basic principles

The key to unlocking the full potential of your home is to work with it, not against it. While it is possible to transform any home almost beyond recognition, you can't make a cottage out of a palace, or a farm out of a high-rise flat. However, with the help of crystals, you can warm up a cold room, shed light on a dark corner, enhance the security of your home and much more.

Fire opals from Mexico, faceted, from palest lemon-white to rich orange.

The appropriate stones can be placed in the affected part of your home (as a focus to help bring about the change you seek) or carried as talismans.

For example, if you wish to create more space you could put an aqua aura in your pocket or purse. Every time you notice it there, it will prompt you to reflect on your need, so that by being reminded of it in all sorts of places, at different times, a novel solution may spring to mind. On the other hand, by positioning the stone precisely in the place where your need is greatest it can spur you to look at your home in a different light. You might see a way to rearrange your belongings or activities in order to give greater priority – and space – to one particularly important area.

If you need some inspiration on how you might redecorate a room, try taking a piece of opal with you when you go looking for paint, fabric or furnishing swatches, or place it in a prominent position in the room, where it may help to inspire you.

HARMONIZING YOUR HOME

To bring happiness and harmony into your home, it helps to explore its various components – somewhere to cook, enjoy recreational activities, bathe and sleep – in the light of an ancient formula: the elements.

The four alchemical elements of fire, air, water and earth have been feted in the Western world for some 2,500 years as containing everything necessary for life. Also, when they are properly balanced in their correct proportions, they stimulate physical wellbeing and promote mental harmony and health. An attractive way to accomplish this balancing act is by using crystals to enhance the appropriate elements of the rooms in your home.

This ancient formula also contains a fifth element, the spirit, or 'quintessence', which is nurtured by, and brings life to, the four elements. In our analysis of the home, the spirit represents you, the person who inhabits the material world composed of the four elements.

THE ELEMENTS IN YOUR HOME

Element	FIRE	AIR	WATER	EARTH
Associated with	Summer Noon South Red Hot and dry	Spring Dawn East Yellow/purple Hot and wet	Autumn Dusk West Blue Cold and wet	Winter Night North Green Cold and dry
Room	KITCHEN The kitchen is the heart of the home, and is characterized by the heat of cooking.	LIVING ROOM A place for recreational and social activities, entertainment and study.	BATHROOM Cleansing the body, calming the emotions and refreshing ourselves, here we also tend to wounds and soothe aches and pains.	BEDROOM A place for rest, sleep and dreams.
Suggestions	Colourful transparent crystals hung in windows. Red veined ornamental stones, such as serpentine, make stylish candleholders. Gem sculptures of animals. Fossil teeth and skulls.	Slices taken from large agate geodes make attractive wind chimes. Crystal mobiles. Transparent crystal spheres or clusters on windowsills. Gem sculptures of birds or fossil insects in amber, for example.	A bowl of tumble-polished gems. An indoor waterfall filled with gems. Gem sculptures of fish, fossil fish and seashells.	Cactus plants with tiny crystals as a gravel topsoil. Pretty jewellery and storage boxes carved from opaque gems. Gem sculptures of plants. Fossil plants, such as petrified wood.
Associated gems	carnelian red amber jasper rose quartz garnet ruby	amethyst iron pyrites ametrine fluorite kunzite tanzanite	sodalite turquoise azurite blue lace agate aquamarine sapphire	aventurine malachite jade unakite peridot emerald
	carnelian	*amethyst*	*sodalite*	*aventurine*

Using crystals with feng shui

At the heart of the Chinese tradition of *feng shui* lies the notion of flowing. The words *feng shui* actually translate to mean 'wind and water' – powerful elements that can be destructive when in full force, yet they can also be as gentle as a summer breeze. *Feng shui* is the art of working with the flow of another force, known as *chi* – the life energy that permeates the entire universe.

UNDERSTANDING CHI

Chi, the ancient sages proposed, is responsible for the wellbeing of every living thing. Like the wind, if you have too much you may be swept headlong into disaster; like water, if you have too little you will grow weak and sick. But in suitable proportions *chi* is thought to bring both health and happiness. As ancient scholars sought to channel and

ACHIEVING A BALANCE

Yin		Yang	
Attributes	*Gems*	*Attributes*	*Gems*
empty	dull	full	sparkling
cold	rounded	hot	angular
dark	black	bright	white
relaxed	amorphous	energetic	crystalline
diffused	composite	focused	pure
passive	opaque	active	transparent
submissive	soft	dominant	hard
sleepy	heavy	alert	light

harness its power for the benefit of mankind, today their Eastern ideas and philosophy have captured the Western imagination, since common sense suggests many of their theories are wise.

BALANCING YIN AND YANG

In order for *chi* to flow, it is necessary to create a space for this to occur – somewhere empty and somewhere full. These extremes, called *yin* and *yang* respectively, are perpetually merging as the *chi* flows on, and their existence sums up the eternal mystery of the universe.

By recognizing the qualities of *yin* and *yang* in everything around us, we can begin to understand the forces that govern the flow of *chi*. Then, if we see too much of one extreme, we may reduce its impact by introducing more of the other. If there's too much *yin*, balance it with some *yang*, and vice versa. Using the chart (left) you can choose which type of crystal to use to correct any imbalance. For example, if a room feels empty, cold and dark – all *yin* attributes – then introduce some *yang* by adding sparkling, angular, white gems.

USING THE CRYSTAL 'BA GUA'

The crystal 'ba gua' (right) is a modern interpretation of the *ba gua*, the *feng shui* geomancers' map with which your home's different areas are calculated. If the rooms in your house don't correspond exactly with the uses suggested by the chart, simply overlay it

THE CRYSTAL BA GUA

NORTH
career
vocation
progression
carnelian
garnet
ruby

ruby

hematite

NORTHWEST
new beginings
helpful people
travel
rock crystal wand,
cluster or
double terminator

rock crystal

NORTHEAST
wisdom
knowledge
enlightenment
hematite
black onyx
smoky quartz

WEST
creative energy
fertility
pleasure
pearl
white opal
moonstone

EAST
family
children
health
lepidolite
black opal
snowflake obsidian
banded fluorite

banded fluorite

moonstone

SOUTHWEST
marriage and
relationships
happiness
peace
malachite
bloodstone
emerald

SOUTHEAST
wealth and prosperity
possessions
lapis lazuli
turquoise
sapphire

sapphire

SOUTH
fame
publicity
reputation
tiger's eye
topaz
diamond

malachite

tiger's eye

on a ground plan of your home to see which stones will be most beneficial in each area. For example, you could place a piece of hematite on a north-east wall, perhaps by sticking small pieces onto a mirror or picture frame, to boost education; or, to rekindle a relationship, try keeping malachite or bloodstone in the south-west area. Using harmonic stones in your home means that the circulating *chi* finds resonance and boosts the aspect of life represented on the *ba gua*. The following pages provide a range of project suggestions for every room, right from the entrance way to the bathroom and office.

The way in

The main functions of the entrance to your property are to provide a sense of homecoming for you and your family, a warm welcome for your friends and relatives, a pleasant view for your neighbours and the local community and a barrier and deterrent to unwelcome strangers. There are a variety of ways in which crystals can have an effect on everyone who crosses your threshold. (If you have a front garden, see pages 46–47 for more ideas and suggestions.)

GUARDING YOUR HOME WITH SENTINELS

The sentinels are a pair of stones that should be positioned on either side of the front path, near the entrance to your property. Resembling the megalithic monuments of some British Bronze Age stone rows, they represent the opposites in nature, such as dark and light, winter and summer, woman and man, the unknown and the known, *yin* and *yang*, and so on.

As you walk between the sentinels you pass through the point of balance at their centre and become the focus of life's harmonized energies. The stabilizing effect is two-fold, promoting mental equilibrium and a sense of proportion on the outward journey (where untold adventures, trials and temptations may await), and then again on your return, when the stresses of the day may be neutralized and left behind you.

The sentinels are powerful guards that play an important role in protecting your property. In addition to the calming effect they will exercise on you, they may also exert a subliminal pressure on an unwelcome visitor. For example, they may defuse the aggressive sales technique of a door-to-door hawker or present an obstacle to a would-be intruder.

To further bolster their talismanic energy, place a small pebble (ideally a sphere) of a black stone, such as obsidian, under the dark sentinel, and a similar orb of white stone (marble, perhaps) beneath the light sentinel.

Depending on the size of your garden and budget, the sentinels may be as small as fists or as tall as men, provided they are both in proportion. The northerly, feminine sentinel should ideally be low, rounded and

Colourful crystals attract chi into your home (from left to right): white howlite, carnelian, hematite, rock crystal, sodalite, tiger's eye, aventurine.

CREATING A RAINBOW PATH

If you have a gravel path or driveway, you could perform the delightful ceremony of scattering a rainbow path.

The path is a symbolic crystal bridge (connecting the world of mortals with the realm of the gods) which links your home to the outside world yet sets it apart in its own realm of peace and plenty.

To create the path you will need a selection of transparent or translucent stones, coloured ones if you like, although you don't have to find stones to represent all seven colours of the rainbow as the name might suggest – the emphasis is on beauty and intuition.

For example, you could choose roughly an equal number of blues, reds and greens of all shades and

The stones on a crystal path provide a way of approaching your home as a new experience: potentially wonderful, always unique and ultimately magical.

tints, or you could just use rock crystal. The smaller the stones, the better – even tiny chippings are very effective – because in this case it is quantity not quality that counts. Once you have gathered a good sample of stones, leave them in a glass bowl on a sunny windowsill until you actually see a rainbow in the sky. At this point, you can dash out and carefully sprinkle the stones all along the pathway.

Like sowing seeds of good fortune and joy, this can be done again and again. If you don't have a gravel path, you can have just as much fun and gain the same effect by scattering the stones among the hard core or concrete while laying a solid path or driveway.

broad. A dark-coloured stone is best but a sedimentary rock, such as limestone, would also be fine. Conversely, the masculine, southerly sentinel should be upright, tall and pointed. A light-coloured stone is best, and an igneous rock, such as granite, is preferable. (If your path falls directly due north or south, then position the north sentinel to the west, and the south sentinel to the east.)

The sentinels can be integrated into your garden in many ways. You could allow moss to take hold or small plants to take root in crevices or pockets where a little soil will provide sustenance, particularly on the feminine sentinel. If you wish to grow flowers around them, you can plant a range of dark-coloured, purple, deep blue or greenish blooms close to the dark sentinel and, in contrast, light-coloured, red, orange, yellow or white flowers close to the light sentinel. Such consideration will enhance and boost their powers.

The entrance, hallway and stairs

The main entrance to your home is the door that everyone tends to use most frequently, and it is where you, your friends, and traditionally your luck, are welcomed inside. A warm, south-facing entrance will benefit from a touch of the elements of air or water (see the chart, Achieving a balance, page 26) to balance any feelings of hot oppressiveness that you may notice, while a north-facing entrance may require the energizing element of fire to bring some light and warmth into a cold area.

A dragon's eye acts as a powerful guardian to watch over the threshold

Large specimens, such as a weighty crystal cluster, or gem displays create a memorable first impression in the hall.

If you have ever experienced difficulty in rejecting the hard-sell techniques of door-to-door salesmen, keep a dragon's eye near the front door. This acts as a powerful talismanic guardian to watch over the threshold and helps you to keep undesirable guests at bay.

If the main door opens directly into a room, such as in a bedsit, you may wish to fashion a divider to create a kind of small hallway. Rather like placing a rock in the midst of flowing water, this helps to prevent *chi* energy from running directly into the room and encourages it to circulate. Using a bookcase is perhaps the easiest means, providing shelves for minerals, candles and plants.

Feng shui principles advise that the front door should not be directly opposite a back door because *chi* could flow in through the front and straight out at the back, bypassing the main part of your home altogether.

Nor should the main door be directly opposite the stairs, since any *chi* accumulated in the upper storey could flow down the stairs and straight out of the front door. The easiest way to overcome these problems is to hang a gem mobile or wind chime (or a combination of both) from the ceiling just inside the door. A spiral pattern is most beneficial: it encourages *chi's* natural curving, flowing lines to spread throughout the home.

Faceted 'crystal' glass is useful here, and has the additional benefit of scattering rainbows around the hall and/or stairway to create an appealing, entrance to your home.

STAIRWAYS

Stairways are often fairly dark places and can seem forbidding. They can also be focal point for accidents in the home, so it is essential that they are kept clear of clutter and distracting light play and shadow.

Bearing in mind the importance of safety, there are a range of options to make stairways more cheerful and to encourage *chi* to circulate upstairs. Small rock crystal wands or delicate window ornaments with facetted glass can be hung in any windows, or you can place rock crystal clusters on the windowsill. If there are no windows, try positioning a mirror to illuminate a dark corner and hang a sparkling gem mobile in front of it to reflect and maximize the light.

HALLWAYS

Hallways are the thoroughfares of your home and because you don't usually linger in them (unless your telephone is in the hallway) they can seem rather neglected and impersonal.

However, as any true pilgrim will testify, the act of travelling is as important as the

A dish piled with a selection of small, tumble-polished gems, placed where the light falls upon them, is ideal in the hallway.

destination, so why not use the space offered by a bare wall for a display cabinet or shelving? This will provide the ideal opportunity to showcase a colourful variety of crystals and gems, particularly those that are better left untouched, such as fragile sulphur, or even cinnabar, which is beautiful but poisonous.

The living room

The lounge or living room is where friends are ushered upon their arrival and where the family gathers to enjoy leisure time. Often a busy, noisy area, where many different activities can be happening at once, the living room is traditionally connected to the element of air, and is often filled with electronic equipment such as television, video and stereo systems.

Where possible the room should feel light and spacious, with furniture positioned around the side walls to leave the central area free. This encourages the free flow of *chi* and makes visitors feel more comfortable: this is because there's less risk of anyone creeping up behind them.

Generally, cool blue or green colours are advisable for a hot south-facing room, and warm red/yellow colours for a cooler north-facing one. There are exceptions, however. A south-facing room overlooked by a hill or tall buildings can feel cold, regardless of how much sunlight actually enters, and the barrier can also obstruct the flow of *chi*. In this situation, hot colours and fiery stones such as topaz, yellow amber or tiger's eye can be a very effective way of introducing warmth. The soft, soothing glow of lit candles can also work wonders.

When choosing stones for this communal area, make sure you include rock crystal, the archetypal gem for air, and place the stone (ideally a sphere) where the sunlight can shine through it. Large specimens can be prohibitively expensive but size is not an

Warm red and yellow stones, such as amber (top) and red jasper (centre, left) will help to warm up a cold, north-facing room, while blue and green stones, like green malachite (centre, right) and spectrolite (right) will help to cool down a hot, south-facing one.

issue here. A small sphere with internal flaws will fill with irridescent rainbows and is actually more effective at promoting the free flow of communication than a large, clear stone.

A particularly attractive idea for the living room is to hang beaded curtains in a doorway or at the windows. Clear, light purple, or golden-coloured gems are best, but beware if there are small children around – the curtains should be hung well out of their reach.

It is possible to buy large quantities of crystal beads (such as rock crystal, amethyst and citrine) and make the curtains yourself. Even small gems, however, can be very heavy in large numbers, so if such a project

appeals, make sure you use the strongest threading fibre available. These strings should be attached to a sturdy wooden or metal rod that can be fixed to the door or window frame by strong brackets.

To ensure that the curtain does not become a barrier to the circulation of *chi*, position the strings fairly widely apart. Remember, the curtains are there to enhance the quality of your home, not to exclude draughts!

Displayed in a glass vase, this selection of tumble-polished eggs, spheres, natural crystals and fossils can help boost the flow of chi *energy in the living room.*

MAKING A MINIATURE LANDSCAPE

Depending on how large or small you wish to make it, this attractive miniature landscape could be a perfect centrepiece for a dinner party or a simple decoration for the mantelpiece. Based on *feng shui* principles and incorporating the *yin* emblem of a lake and the *yang* emblem of a mountain, it will help to create a harmonious atmosphere at any social gathering. You could also make a permanent decoration by substituting tumble-polished milky quartz stones or moonstones for the water in the bowl.

Fill a large, circular dish almost to the top with fine sand to form the setting for the landscape. Place a bowl of water midway between the edge and the centre and then drop a handful of red stones into the bowl (do not use fire opal because it can absorb water and crack). Smooth the surface of the sand and place a black pebble similar in size to the bowl at the midway point, opposite the water. Position the dish so that the mountain is to the north and the lake to the south. Place a white stone to the west, closer to the water than to the mountain. Place a blue stone to the east, closer to the mountain than to the water. To finish, decorate the sand by drawing fine lines or, alternatively, sprinkle tiny sparkling stones on the surface.

This feng shui *landscape is designed to ensure a happy atmosphere between friends and family at dinners and parties.*

The kitchen

The kitchen is often regarded as the heart of the home and so should always feel welcoming. When food is being prepared and cooked, it can get very hot, although as the kitchen is often situated at the cooler side of the house the room itself can sometimes feel cold. To encourage a calm and comfortably warm atmosphere, place a polished, rosy pink, opaque rhodochrosite on the windowsill. Pink minerals are associated with physical love and affection and epitomize the care that goes into feeding and nurturing the family.

To further balance the *yang* influence, place a bloodstone, with more green than red, beside the rhodochrosite. Green stones symbolize maternal love and remind us to take special care in the kitchen, where the majority of domestic accidents occur.

Alternatively, hang a piece of watermelon tourmaline in the window so that light can shine through it. This translucent gem contains both pink and green hues which represents both *yin* and *yang* in one unified whole, which helps create a feeling of unity within the family.

It is best to avoid using crystal mobiles in the kitchen since grease particles in the air quickly dull and cloud transparent stones and cleaning them can become rather tedious. To help to distract you from mundane tasks keep a strongly chatoyant mineral, such as tiger's eye, on a shelf or at eye level so that you can see it as you carry out your chores.

THE POWER OF SALT

One gem that almost everyone will have somewhere in the kitchen is salt. Referred to as halite by geologists, some grains of this household rock show facets of its cubic crystalline structure.

Salt has been a tradable commodity since ancient times and rock salt is generally mined from deposits left by evaporated primeval seas or the saline waters of inland lakes (sea water contains about 2.7 per cent dissolved salt by weight). These deposits are still forming, for example at Great Salt Lake, Utah, in the United States. Places associated with salt mines were often named accordingly – there are many such towns in Britain and more around the world, among which Salzburg, Austria, is particularly well-known for supplying ancient Rome.

People with surnames such as Salter probably had medieval ancestors who earned their living from the mining, transportation or sale of salt, which was once used in huge quantities to preserve food. Even the word 'salary', from the Latin sal for salt, derives from the name given to an allowance paid to Roman legionnaires so that they could purchase salt to replace what they lost in the toil and sweat of their duties.

This edible mineral is symbolic of cleanliness, purification, human worth and, of course, the sea from which all life came. Keep it on view as a constant reminder of life itself, even if you choose not to display any other crystals in the kitchen.

As well as being useful for cooking, rock salt is a reminder of simple beauty. It symbolizes cleanliness and purification and should be kept on prominent display in the kitchen.

Pink minerals, such as rose quartz, are linked with physical love and affection, and the nurturing associated with feeding others

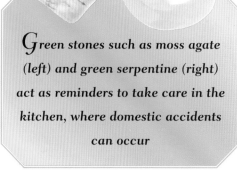

Green stones such as moss agate (left) and green serpentine (right) act as reminders to take care in the kitchen, where domestic accidents can occur

The dining room

Often the kitchen serves as both a cooking and dining area, which is not ideal in some respects, because the kitchen is symbolic of the element of fire while the dining area should be more earth-orientated. However, there are ways to achieve a more harmonious ambience, such as to set the dining table cosily in a corner rather than in the middle of the room. This avoids obstructing the flow of *chi* and helps people sitting at the table (on the wallward sides anyway!) feel secure – which in turn aids the digestion and promotes happy conversation.

Opal is a feast for the eyes and a reminder of the richness of the dining table

Crockery and glassware can be chosen to suit the occasion and to create the right environment. Solid, earthy, pottery and clayware, for example, is perfect for a serious heart-to-heart with a friend because it reinforces the solidity of the relationship. A more frivolous get-together would be better served by glass tumblers and bamboo or woven serving platters. A rock crystal mobile – or even a crystal chandelier if your dining room and budget can take it – can be hung above the dining table to aid the flow of *chi* and help keep the conversation buzzing.

A ROMANTIC DINNER

For a romantic dinner, you can choose two candles of contrasting colours – warm rose red, and green or turquoise is a good combination – and glass or crystal candle-

A crystal landscape for your dining table creates a feeling of intimacy and calm that is perfect for a romantic dinner.

holders: position them so that they form a portal through which you can gaze at each other. Around the red candle's holder, place a selection of blue and green stones: chrysocolla, turquoise, aventurine. Around the other candle's holder, heap warm coloured stones: rose quartz, red amber, citrine. Between the candles and slightly towards the woman's place setting, position a rock crystal wand or a small jade obelisk or pyramid (a symbol of protection and the quest for knowledge). On the man's side, place either a rock crystal sphere, or a small jade sphere (symbolizing eternity and limitless potential). Rock crystal represents communication, while jade symbolizes love: either is appropriate here.

This arrangement carries several benefits. Firstly the stones will help to earth an emotionally charged atmosphere, so that you don't get carried away against your better judgement. They will also act as reservoirs of bonhomie, smoothing over any potential sticky moments. But most of all they promote communication, both verbal and in terms of body language.

Crystals can help create atmosphere; try warm reds for an intimate supper, or translucent amethyst for a sparkling dinner party.

AN INFORMAL LUNCH WITH FRIENDS

The emphasis for this sort of meal should be on easy conversation, laughter and fun.

Ideally, a bowl or chunky dish full of fairly large stones, placed among the food dishes, forms both a talking point and a reminder that nourishment does not consist entirely of

You can display jade to bring peace, prosperity and love

edibles – food for the spirit plays a significant part in life too. Choose robust, non-porous stones so your friends can touch and handle them without harming them. Try serpentine worry eggs, large pieces of rock crystal, and hematite spheres.

A BUSINESS LUNCH

An attractive crystal centrepiece will encourage your lunch to run smoothly and successfully.

A variation on the *feng shui* principle of the financially beneficial tank of goldfish is to position in the centre of the table a clear glass bowl of water, containing as many yellow and gold-coloured stones as you wish (uneven numbers are especially lucky, remember). Tiger's eye is excellent for confidence and magnanimity, citrine for eloquence, rutilated quartz to help you stay in touch with the modern world. For extra luck, float three or five gold candles in the bowl.

The bedroom

As we spend, on average, approximately a third of our lives asleep, the bedroom deserves considerable attention. This room has four main purposes: it is a place for loving and the intimate expression of affection and passion; it is a place for sleep, where we recharge our batteries; it is where we dream, when our subconscious minds can order the day's events and summon the future; and it is where we wake, ideally refreshed after a sound night's sleep.

Keep a bag of gems under the pillow or near the bed to counter nightmares and help you sleep.

According to *feng shui* principles, the bedroom should have as few internal corners as possible and the bed should be diagonally opposite the door. It may also be of benefit to align the bed on a north/south axis, so that the bedhead is against a northerly wall. To neutralize a corner, place an artificial plant or a substantial stone on the floor there (provided you won't trip over it) or hang a crystal from the corner of the ceiling.

A screen, or room divider or even a chest between the bed and the door encourages the

Darker gems in purple and red, such as amethyst (centre), fluorite (right), and carnelian (left), along with black gems, can boost sensuous bedroom energy

circulation of *chi*. But note that the foot of the bed should never face the door. This area also creates a display point for plants and beneficial stones. Try hanging a crystal mobile just inside the door, or a crystal over the bed to help you sleep.

CREATING A MOOD

Crystals can enhance many different ambiances and you can choose them according to the mood you wish to create.

The bedroom may be a focus for intimacy, in which case you should display the darker opaque stones in shades of purple, red and black, preferably with swirling patterns on their surfaces. Charoite, sardonyx, falcon's eye, lapis lazuli and tiger iron are all ideal. Carved figures and statuettes are also perfect here.

These can range from dragons carved from tiger's eye, jade or amethyst, to simple animal shapes inherent in the markings of the gem itself. Spheres and egg shapes, representing fertility and femininity are also good, together with an upright crystal wand to balance them. Some people, even if they loathe snakes, may feel deliciously wicked in the presence of a jewelled cobra, while others might see a dragon's eye pig as a symbol of luxury and fun.

DREAM TIME

There are certain stones that may aid restful sleep. For example, to help understand your dreams, keep a double-terminated rock crystal close by while you sleep. To promote a deep, healing sleep, try putting a tumble-polished piece of hematite under your pillow.

To help counter nightmares, place a large, smooth piece of moss agate or tektite by the bed where you can touch it. Its solid, reassuring feel will calm you if your sleep is broken.

If insomnia is a problem, try gazing into a small sphere or tumbled-shape of green aventurine. Allow its soft, gentle hue, twinkling spangles of mica and smooth curves to lull you to sleep. Finally, if you use an alarm to wake you, keep a chunk of clear, yellow amber near the clock. As you switch off the alarm, pick up the stone. Holding something akin to solid sunlight in your hand before getting up can make all the difference to the quality of your day ahead!

Decorate a bedside lamp with a few of your favourite gems, or those that will help promote a sound night's sleep.

Children's rooms

Children's rooms are often places to play as well as sleep. During the day, it is important that they are as light as possible. Crystal mobiles hung in windows cast all the colours of the rainbow into the darkest corners of the room and provide a source of fascination and delight for children of all ages.

When playing, children benefit from being stretched in new directions and, if possible, the play area should stimulate all the senses. Using basic *feng shui* principles, shiny crystal mobiles and tinkling wind chimes will both encourage the flow of *chi* and provide a focus of interest.

An amber bear, glowing in the sunshine, makes an appealing ornament. To avoid accidents, it is best kept out of reach on a higher table or shelf.

INTRODUCING CHILDREN TO STONES

When buying crystals for children, it is best to opt for unpolished, unadulterated specimens because they help youngsters to appreciate and understand where crystals come from and how they look in their natural state.

Children are always on the look out for something new and interesting. Crystals and other minerals offer diversity and natural beauty, and they can provide endless amusement as well as being educational. To encourage children's interest and use of stones, you could take them shopping to choose a handful of gems for themselves. Pay attention to their choices, and if they are notably biased toward one colour or shape, buy a few of the others – 'for yourself' – while making a point of allowing your child to share or borrow them. Create a magical place for your child to store the stones: a simple bowl will do, although a miniature treasure chest might be more fun!

Name the stones, make up stories about them and try to find out about their history, what they can be used for, and where they can be found. In other words, balance the fanciful with the real to help children expand their worlds. Allow them to collect stones from the garden, park or beach, wash them and examine them, so that they can learn about the geology beneath their feet.

Crystal clusters are particularly good for children as they enable young minds to grasp the idea that although all the various individual crystals are the same mineral (and emanate from the same source), their sizes, shapes and other details are individual and unique. This simple lesson can promote an understanding of the diversity of life.

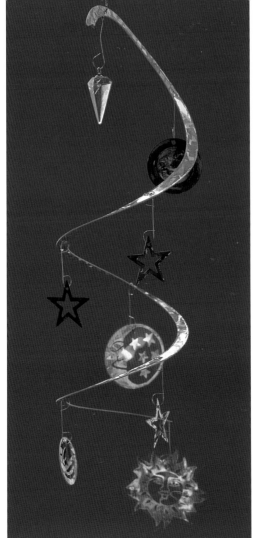

A sparkling crystal mobile provides a source of endless fascination for children of all ages.

THE COLOURS OF CHILDHOOD

Have you ever wondered why it is traditional for girls to be dressed in warmer colours, such as pink and red, and boys in the cooler shades, such as blue? In *feng shui* terms, the female side of nature is seen as *yin* – or cool, dark, submissive and receptive – and is symbolized by the colour blue. The male side of nature is thought of as *yang* – hot, bright, dominant and active, which is embodied in the colour red.

To compensate for too much *yin* or *yang*, you can counteract one with the other. So, to balance a girl's innate *yin*, decorate her room with *yang* pinks and reds, and to counter-balance a boy's natural *yang*, use cool shades. This might not be to your child's taste, but you can usually compromise. You could try substituting warm purples or orange hues for pinks and reds and use gems such as amber, amethyst and carnelian.

STONES AND SAFETY

It is vital to be careful when handling stones around very young children and babies because many crystals, such as quartz, are brittle, liable to chip and may break if dropped on a hard floor, causing dangerously sharp shards to scatter everywhere. Another potential danger is that small, colourful stones could be mistaken for sweets, and might easily be swallowed or worse, cause choking. However, with sensible precautions and parental supervision, the beauty of these stones can stimulate infants in a way that is wonderful to share.

Blue and pink stones can be used in children's rooms to balance yin (female) or yang (male): pink to warm cool yin, and vice versa

sodalite

ruby

blue howlite

rose quartz

The bathroom

A place for calm, meditation, fun and pampering, the bathroom is associated with the ancient tradition of spas which have been used for thousands of years by those in search of spiritual inspiration, relief from physical ailments and an escape from the cares of the world.

Yet bathrooms can be rather basic and functional – they're often alternately hot and steamy or cold and clammy. To make the bathroom more inviting, be bold and adventurous with your colour schemes, display plenty of fluffy, brightly coloured towels and put up shelves and display areas to show off crystals, plants and bottles of bath oil.

Pumice stones are renowned as healing stones and will gently smooth away hard skin.

Transparent or translucent, faceted, large crystals are best for the bathroom, as they won't be dulled by the steam. Display bath salts, too, as these actually contain crystalline mineral salts. You could also display tiny gemstones in bottles, and hang rock crystal wands from the ceiling and shower-curtain rail. Try draping faceted crystals from wall sconces that hold candles. The lit candles will bring the element of fire into a room that inevitably contains so much water.

Keep a real pumice stone handy (which is a healing stone in itself) both to smooth away hard skin and to remind you that not all beneficial stones are sparkling, crystalline display pieces! You can also use clay as a face pack to invigorate your skin.

Soak in the bath with your favourite healing stones, and keep some massage stones within easy reach when you climb out of the tub. The best massage stones are large, smooth and rounded or smooth-ended wands, which you can use to gently rub and stroke the body.

Taking a bath can be a sensual experience or it can simply be fun – sprinkling a handful of bath salts into the water will liven up bathtime for children, who love to watch the way in which they fizz and dissolve, turning the water a different colour.

You can use fragrant aromatherapy oils in the bathroom to make it smell fresh, flowery, exotic or fruity. And if you're feeling particularly stressed or tired, you could add a crystal essence remedy to the bathwater to soothe away the strains of the day, or refresh and revive the body (see page 72 for details on making your own remedy).

Celestite adds an air of tranquillity, helping relaxation

Keep a selection of massage stones handy so that you can treat yourself or your partner to a relaxing rub-down.

The garden

Whether it is a small window box on the sill outside the kitchen or a rambling country estate with acres to spare, the garden is an ideal place for stones of all shapes and sizes. Try to channel *chi* energy into your garden by creating a water feature (preferably to the south of the site), with a fountain or waterfall to keep *chi* moving. Surround it with large stones, perhaps containing mica flakes that sparkle in the light.

Planting a small aventurine stone with your seeds can encourage them to grow

Arrange stones or shrubs along pathways to guide *chi* around the entire site, soften corners with climbing plants, or better still, trees, and hang crystal mobiles from the branches. It may be possible to hang aolian chimes too, adding an element of soft music to the garden.

DRY LANDSCAPE GARDENS

Often called Zen gardens, Japanese (or Chinese) rock gardens, these contain no plants at all, only stones. In their simplest form they are just large rocks set on a bed of gravel.

The Chinese term for landscape is 'mountain water'. The gravel bed can be of any size and shape: straightforward rectangles are fine. The aim is to create a peaceful place that provides an opportunity for meditation.

Achieving harmony between *yin* and *yang* rocks is important. Smooth, rounded, or dark specimens for example, need to be balanced by rough, angular or light-hued rocks. But beware, as traditionally it would be a very brave

Use crystals to make a focal point of a favourite container plant on the patio.

USEFUL GARDEN STONES

Whether you use your garden to cultivate fruit and vegtables or simply to relax and unwind, there are stones to suit the purpose. Here are some beneficial stones for a number of different requirements:

- **To grow healthy fruit and veg**
Try 'planting' a small stone with seeds as a talisman to encourage them to grow. Malachite and aventurine, because of their association with fertility, are particularly good for promoting healthy, strong growth.

aventurine

- **To boost the power of herbs**
Place a piece of hematite, smoky quartz or basalt at the base of each plant for strength and fruitfulness.

hematite

- **To promote rest and relaxation**
Choose large stones, such as ammonites and fossil-bearing rocks, and position them close to your favourite seats; hang sparkling mobiles of prehnite, tourmaline or amethyst from the branches of trees and bushes.

amethyst

- **To enhance the view**
Chalcedony, smoky quartz, chrysoprase and dumortite look beautiful when bathed in sunlight because of their contrasts in colour and texture.

chrysoprase

- **To create a peaceful atmosphere**
Rock crystals, amethyst spheres, labradolite and jade are all excellent stones for aiding tranquillity and meditation.

rock crystal wand

---

gardener who places an even number of rocks in the garden, as this is thought to be unlucky.

Choosing the number and character of rocks and finding their best position on the gravel is an art that anyone can master. Give yourself time to live with a pattern: walk around it, study it from all sides, and watch how the creeping shadows affect the layout at different times of the day – and year – when considering a new arrangement of garden stones.

EARTHING STONE

An earthing stone acts as a point of contact with the earth deep beneath your feet. Like a

Crystals can help cacti enjoy a dry, exotic landscape.

lightning rod, it can conduct your thoughts to the ancient body of the earth, acting as a channel through which you may download all the conflicting energies which we pick up during our stressful lives.

This stone is arguably the most powerful stone you can use for cleansing, balancing and attuning yourself with the nature around you. It doesn't need to be large, provided it is positioned in a place of its own where it can radiate a sense of calmness. It must be in direct contact with the soil and should ideally be made of the same sort of rock as the bedrock below the soil in your own plot.

The office

Because work takes up a considerable part of our time, it is very important that our working environment feels pleasing to the eye and is also conducive to our mental wellbeing. If you work from home, for example, you may find that you have to adapt part of an existing room to make space for your office. But wherever you intend to work, there are a number of basic *feng shui* principles to follow when arranging your equipment.

A heavy stone, such as this hefty chunk of iron pyrites, makes an impressive office paperweight.

Try to position your work station, whether it's a desk, worktop or sewing machine, towards the east in order to benefit from the *yang chi* engendered by the rising sun. If possible, you should have your back to a wall and use a high, solid bookcase or divider to separate the work space from the rest of the room. It may also be a good idea to create some kind of 'official' threshold to your work area, similar to the sentinels at the entrance to the house (see pages 28–29). Indoors, you could use one tall and one short potted plant, or a couple of tables, chests or filing cabinets with crystal clusters prominently placed on top.

During the day, you can keep several favourite stones at hand, especially if your work involves staring at a computer screen for hours at a time. Contrasting shapes, colours and textures are ideal, for example a rock crystal wand, a large smooth piece of carnelian, or – a particular favourite – a smooth piece of polished labradorite.

To give your mind, eyes and fingers a break, devote a few minutes every hour to handling your stones and admiring the play of light, especially on stones such as spectrolite, opal and moonstone and, of course, rock crystal with its characteristic 'rainbows'.

CAREER CRYSTALS

As well as promoting concentration and relieving stress at work, stones can have many practical and attractive uses in the workplace:

- *Paperweights*

Natural stones, such as smooth pebbles, or stones carved by hand into all sorts of forms, such as animals or fish, make ideal paperweights. Avoid shapes with sharp edges, though, as these could dent or tear paper.

- *Bookends*

Natural or hand-crafted, heavy stones can be used to prop up books or files.

CRYSTALS TO REDUCE STRESS

- **To boost self-confidence**

 Bright, sunny yellow stones, such as topaz, **citrine** or tiger's eye, symbolize the sun and represent courage and vitality.

- **To aid communication**

 When dealing with colleagues, white and pearly stones, such as **moonstone** help to foster empathy. Various forms of rock crystal may also be useful, as this is associated with mental agility and quick-thinking.

- **For growth and expansion**

 Green stones, such as **chryso-prase**, are associated with fertility and stability, and may be helpful during times of change.

- **For a competitive edge**

 Bright red stones, such as **ruby,** represent straightforwardness and strength in the face of hardship and adversity.

- **In legal and financial matters** **Lace agate** and

 other blue stones indicate business acumen.

- **To help concentration**

 When facing testing times, black stones like **obsidian** represent patience and fortitude.

- **For inspiration**

 To help you come up with new ideas use purple stones, such as **fluorite**, which are associated with dynamism and unconventiality.

- **To foster compassion**

 When sympathy and gentleness are required, soft turquoise-blue stones, such as **black opal**, may help. They symbolize sensitivity and perception.

- **To boost critical abilities**

Multi-coloured stones, like **spectrolite**, represent self-analysis.

- *Desk tidy*

Carved bowls, dishes, vases and lidded pots are all handy for keeping paperclips, pins and pens in order. (For smokers who want to kick the habit, using a crystal bowl as an ashtray may prove helpful, as sullying this delicate object with stubs can be a potent reminder of how smoking abuses the lungs).

- *Stress buster*

Inserted into the soil around the base of a plant on the desk or windowsill, a rock crystal wand can act as a mental lightning rod, earthing your stress and draining your mind of anxiety and confusion.

For a little extra help in other areas of your working life, choose any stones from the list (left) to keep close to, or on your desk. They will help to boost you and may assist with problem-solving, although they can't take the place of professional advice!

Of course, if you work in an office away from home, many of these suggestions will apply to your desk or work space there, although you may have less choice about where your desk and equipment are situated. It is therefore even more important to try to personalize the area, and crystals can be perfect for doing this – they are easily portable, usually inconspicuous and are very individual items.

Finally, consider having a goldfish tank, with bright gem stones on the bottom, in your work area. The colour and movement of these fish help stimulate *chi* and may help improve your finances!

Problem-solvers

PROBLEM	SOLUTION	PROBLEM	SOLUTION
Room(s) feel cramped and lacking in space	Carry a single aqua aura crystal in your pocket or purse to inspire a novel solution. Alternatively, place the crystal in the affected room(s) to encourage you to see the surroundings in a new light. You may then see a way to rearrange your furniture or belongings to create more room	*You are continually pestered by door-to-door sales people or unwelcome visitors*	Position a pair of sentinels, one light-coloured stone and one dark-coloured stone, on either side of your front path, close to the entrance to your property
You are worried about being burgled or are feeling insecure	Display hematite, a protective gem, near the front door or other entrances	*You find it difficult to reject the hard-sell techniques of door-to-door sales people*	Keep a dragon's eye near the front entrance to act as a powerful talismanic guardian to watch over the threshold
Your home seems cold and damp	Imperial topaz brings the essence of the life-giving hearth into your home, making it feel warmer and more comfortable	*After a stressful day, you tend to bring your problems home*	Create a rainbow path to your front door, as this will help you to forget about the pressures of the outside world when you arrive home
The house is dark and does not get enough natural light	Citrine captures the essence of sparkling sunlight, so will help to brighten gloomy areas	*A warm, south-facing entrance or room feels too hot and oppressive*	Introduce the elements of air and/or water by using stones such as iron pyrites, ametrine amethyst, turquoise or azurite to cool it down
Rooms are stuffy and airless	Place rock crystal in the affected rooms – it epitomizes the element of air and will help to banish staleness	*A cold, north-facing entrance or room feels chilly and miserable*	Use stones such as carnelian, ruby, garnet, red amber, malachite, aventurine or jade, which represent the elements of fire or earth, to warm it up
You want to redecorate but don't know which colours to use	Opal, in its various coloured forms, may help to inspire decorating ideas	*The back door of your home is directly opposite the front door, or the front door is opposite the stairs, going against* feng shui *principles*	Hang a gem mobile or wind chime from the ceiling just inside the front door
You are having trouble with the neighbours	White moonstone placed against an intervening wall helps to ease friction. In a confrontation, hold a tumble-polished rock crystal to calm things down	*The hallway is dark and forbidding*	Hang crystal wands in any windows, or illuminate a dark corner with a mirror, and a sparkling gem mobile in front to reflect the light

PROBLEM	SOLUTION	PROBLEM	SOLUTION
The hallway is dull and impersonal	Erect some display shelving to show off your more fragile crystals, such as sulphur or desert rose.	*You dream a lot but cannot understand what your dreams mean*	Keep a double-terminated rock crystal close by while you sleep to help you interpret your dreams
You are hosting a dinner party and are worried that your guests may not get on!	Make a *feng shui* landscape (see page 34) and use it as a table centre to foster a harmonious atmosphere. Alternatively, place a small sphere of rock crystal on the table, preferably a specimen with rainbow-creating internal flaws, to promote the free flow of conversation. As a more permanent feature, a rock crystal mobile – or even a crystal chandelier – can be hung above the dining table to enliven conversation	*You find it difficult to get out of bed in the mornings*	Keep a chunk of clear, yellow amber on your bedside table and pick up the stone as soon as your alarm clock goes off. Yellow amber looks just like a piece of solid sunlight and can really help to give you a kick start, especially on a gloomy winter's day!
		When preparing food in the kitchen, you often cut or burn yourself	Place a bloodstone, with more green than red in it, on the kitchen windowsill. Green gems symbolize maternal love and will remind you to take extra care in the kitchen to avoid accidents when working
Your bedroom is not the focus for intimacy that you would like it to be	Rekindle passions by displaying darker, opaque gems in shades of purple, red and black, preferably with swirling patterns on their surfaces. Charoite, sardonyx, falcon's eye, lapis lazuli and tiger iron are all ideal, as are carved figures and statuettes	*You loathe mundane kitchen tasks such as washing up or preparing vegetables*	Keep a strongly chatoyant gem, such as tiger's eye, on a shelf or at eye level to help distract you from your tedious chores
You suffer from nightmares	Place a large, smooth piece of moss agate or tektite beside the bed, so you can reach out and feel assured by its weight and form	*You often feel stressed at work*	Insert a rock crystal wand into the soil around the base of a plant on the desk or windowsill. This can help to 'earth' your stress and ease anxiety
You are a restless sleeper, or suffer from insomnia	Place a piece of tumble-polished hematite under your pillow or try gazing into a small sphere or tumble-polished piece of aventurine, focussing on its gentle colour and soft curves to help you drift off to sleep	*When growing fruit and/or vegetables in the garden, the results are disappointing*	Try 'planting' a small stone with seeds as a talisman to encourage healthy growth. Because of their association with fertility you could try malachite and aventurine

PROBLEM	SOLUTION
Your herb garden is not growing as well as it should	Place a piece of hematite, smoky quartz or basalt at the base of each plant to engender strength and fruitfulness in the herbs
Your main entry door opens directly into a main room	If possible, buy or build shelving and display large crystal clusters. If there is insufficient space, a screen or curtain incorporating smaller pieces of translucent stones – rock crystal or amethyst – will help provide privacy while enabling the flow of *chi*
You find it difficult to settle in your lounge – it feels too 'busy'	Handling rose quartz, milky quartz, or a tumble-polished piece of chrysocolla may act to soothe your nerves. For a longer-term solution, place ammonite somewhere prominent, where it may be frequently touched and used as a focus for quiet contemplation
To keep the peace when in-laws (or other hostiles!) visit	Wear or carry rose quartz to help foster warmth, peace and friendliness
You find the view from your home displeasing	Hang sparkling crystals in the windows to distract the eye and bring beauty into the house
The front of your house faces tall, upright structures (pylons, chimneys, a tower block etc)	This can prevent the free flow of *chi*. While it isn't normally possible to change external views, spiral or helical crystal mobiles can help encourage *chi* circulate more positively in the house

PROBLEM	SOLUTION
You find the atmosphere in the house depressing	Use topaz, citrine and amber to bring some sunshine into the house, and hang wind-chimes to add a little music. If you feel depressed about coming back home, try wearing amber jewellery and hang a rock crystal wand above the inside of the main door
You can't seem to attract good luck	If possible, place a sample of your sun stone (see pages 76–127) just inside the door (or even better, attached to the inside of the door itself), where you can touch it as you leave and enter the house
You have had a great deal of bad luck	Keep a large piece of hematite just inside the front door. If you feel you are being dogged by ill-fortune, wear a small piece of hematite jewellery and stroke or handle it any time you feel anxious
You are short of money	*Feng shui* principles advise that gold attracts gold, and suggests keeping goldfish in a pool outside the front door. Alternatively, gold-coloured stones – citrine, iron pyrites, and especially tiger's eye – at the door, or placed on something representing your main source of income (your bank book, your computer, your toolbox etc) may help to generate cash!

PROBLEM	SOLUTION
You want to give up smoking	Burn a little amber incense to clear the smell of tobacco smoke from the house. Then use black opal as a focus for meditation and for developing your will-power. (It may also help to carry a piece of moss agate with you, for use as a worry stone)
You dream of having an extension built, but lack the resources	Plant a small double terminator (or herkimer diamond) in the ground where you wish the extension to be – it may act as a catalyst, helping to make your dream come true
You work from home but have difficulty concentrating	Try to keep a contrasting selection of stones at hand – rock crystal (especially a rock crystal wand), labradorite and carnelian, for example – but don't use your tendency to touch and handle them as an excuse to procrastinate!
You're planning a party and want it to go with a swing!	Scatter small groups of rock crystal around the rooms that are to be used: on shelves, on windowsills, on door jambs. Position them where the light will catch them, to ensure your party sparkles!
You are forced to move from a much-loved home into somewhere new	Place a large rock crystal sphere, where the sun can shine through it or place sodalite or dumorturite at the front door to welcome new life experiences. Keep a bowl of tumble-polished rose quartz, milky quartz, amethyst and hematite beside your bed to help you sleep well and feel at home

PROBLEM	SOLUTION
You feel generally unwell – not exactly ill, but not at your best, either	Use milky quartz as a worry stone, and surround yourself with brightly coloured crystals. Amber is excellent if you are feeling 'low', and chrysocolla may brighten your spirits if you feel tired and listless. Constantly handling a rock crystal or amethyst worry egg can be very beneficial if the problem is stress related
You have a tendency to be lazy	Ideally, get up and do something active! If this is too difficult – for example if you really are tired – try using geodes or crystal clusters as a focus for contemplation and meditation – much more attractive than the TV, and the images that emerge from your subconscious as you gaze at and into the stones may be surprisingly illuminating!
Your neighbours are noisy or nosy!	Tact and diplomacy, and often a sense of humour, are called for in this situation. Wear or carry chrysoprase to foster Libran equilibrium and hold black onyx in your hand in any conversation with your neighbours. Snowflake obsidian may be useful in engendering patience and avoiding confrontation
You feel that your home is full of unhappy memories	Focusing on agate slices, which are effectively slices of geodes, may help you to focus your mind on the source of your unhappy memories and help you deal with the distress that they cause

CRYSTALS FOR HEALTH

THROAT CHAKRA

Many folk tales tell of heroic quests for gems purported to have magical healing powers, yet the beneficial properties of crystals are not merely mythical. Just as homeopathic and herbal remedies are enjoying a resurgence today, crystal healing too is widely practised, with much anecdotal evidence to support its efficacy. Placing crystals on the body's chakras, or spiritual 'power points' is thought to encourage healing, soothing common ailments from headaches and insomnia to a sore throat. You can also make your own crystal essence remedies using the safe stones listed on pages 74–75, and learn to use crystals for visualization, helping to restore your energy and good health.

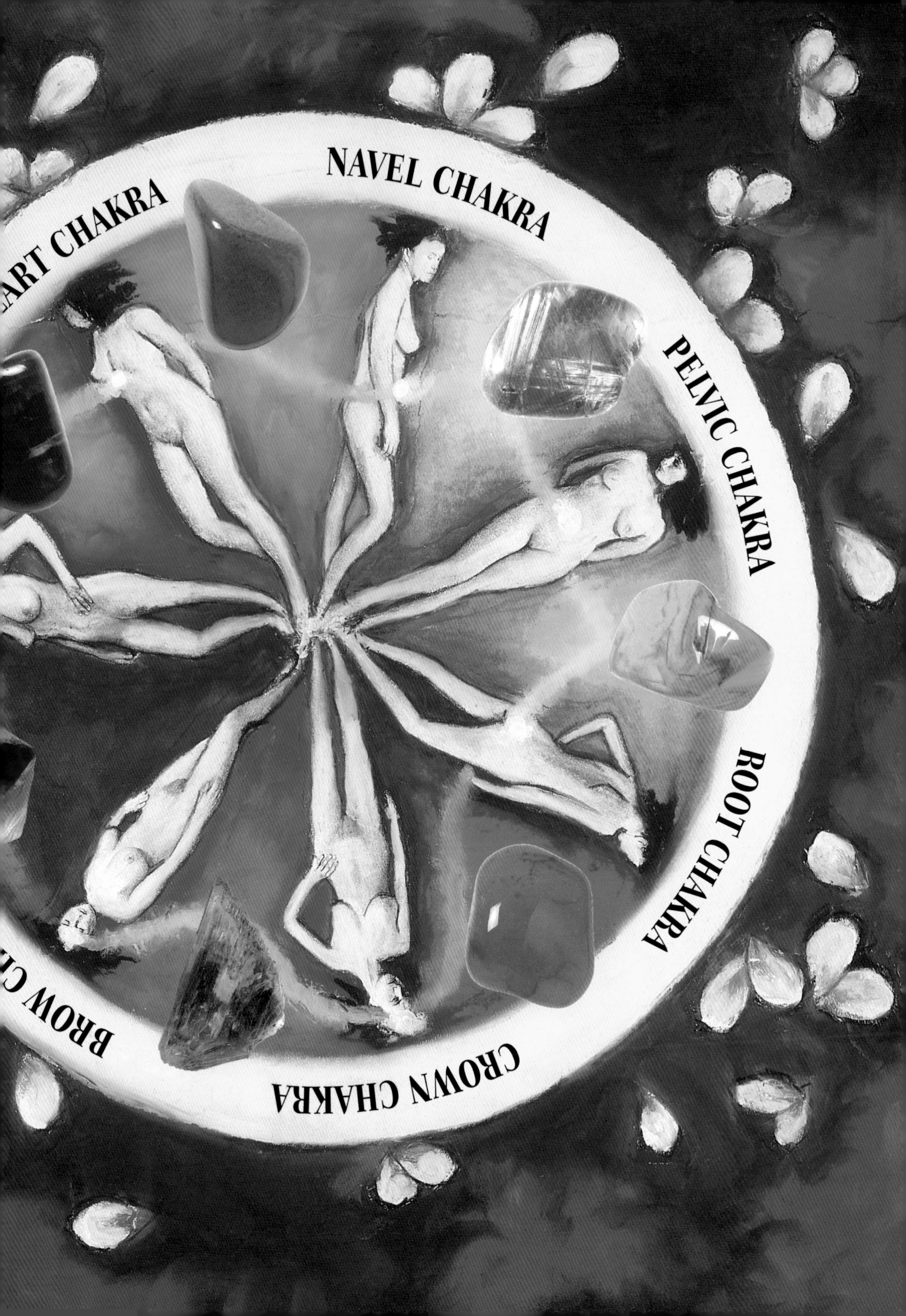

NAVEL CHAKRA
PELVIC CHAKRA
ROOT CHAKRA
CROWN CHAKRA

Crystals and chakras

The revival of ancient therapies

Before modern medicine discovered the viral and bacterial causes of many ailments, the explanations for deadly epidemics and infections were often wildly fanciful. Demonic infestation or curses from angry spirits were common diagnoses and the only treatments for such dreadful plagues were often as dramatic as they were dubious and sometimes horribly self-defeating. The belief that the Black Death of medieval Europe was caused by witchcraft led to the mass killings of witches' familiars – cats – which were controlling the population of rats that spread the disease.

Calcite can help reduce stress, balance the emotions and inspire feelings of calm and relaxation

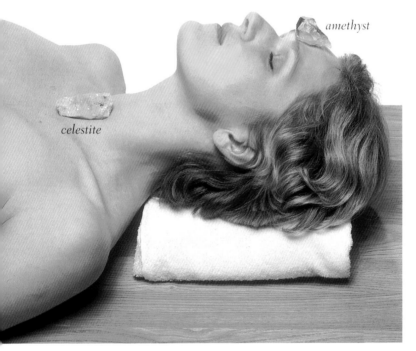

amethyst

celestite

The subtle effects of chakra healing are best felt when you are deeply relaxed.

In the 19th century, pioneers of modern medicine announced the new age of scientific treatment. They discredited the stereotypical image of the witchdoctor, with his magic array of rattles, feather armbands and anklets, necklaces of teeth and sacred stones, to such

an extent that much of their unique herbal knowledge was mocked and ignored.

With the new millennium, many people are becoming more interested in natural methods of healing. Some people may be wary of any side-effects that may be associated with various manufactured drugs; others strongly disapprove of laboratory testing on animals, or are concerned at some of the latest medical experiments in the search for animal organ donors, for example. Traditional remedies, such as herbalism, homeopathy and crystal therapies, can be seen as natural alternatives for, or complementary with modern drugs.

In the very recent swing toward holistic medicine, the value of the witchdoctor's shamanic approach is only now being re-

Some crystal healers treat the lower chakras in an unconventional pattern, and focus on different energies.

Precisely how crystals work to counteract disease or even promote good health is hotly debated. A popular theory is that crystals actually vibrate at the same pitch as people, and can be used to relieve suffering because the resonance between the patient and the stone either combats the vibrations of illness or amplifies those of health.

Protagonists frequently claim that the healing power comes from a 'higher plane', a 'higher level of consciousness' or, quite simply, a 'higher dimension'. However, the problem with such explanations is that because they have no scientific basis, they can be difficult to grasp. Yet it is not always necessary to know exactly how these therapies work before putting them to good use.

A NOTE ON SAFETY

On the following pages, we explain the number of ways in which crystals can be used to help common ailments and maintain optimum health. However, always bear in mind that it is generally accepted that these forms of healing can take several sessions to be effective. If symptoms persist or worry you in any way, always ensure that you consult your doctor.

evaluated and integrated into a wide range of alternative and complementary therapies. There is a broad spectrum of therapies now available which many people have been tempted to try out. Perhaps it is hardly surprising, then, that an increasing number are turning to crystals not only to help cure their ailments but, significantly, to maintain optimum health.

Healing with chakras

The chakras are part of an ancient and complex occult system devised in India and used by Yogis as a path to enlightenment. There are seven primary chakras located along an imaginary vertical line from the top of the head to the base of the spine. These are: *Sahasrara* (crown of the head); *Ajna* (brow, referred to as the 'third eye' chakra); *Visuddha* (throat); *Anahata* (heart); *Manipura* (navel); *Svadishthana* (pelvic) and *Muladhara* (the base or root chakra).

The actual practice of healing with chakras is simply a matter of placing the appropriate stone on the chakra corresponding to the malady. It is believed that the natural affinity between stone and chakra kindles an interaction which retunes the chakra to a healthy vibration, thereby healing the part of the body affected.

YOUR CHAKRA 'POWER POINTS'

Each chakra is generally visualized as a wheel with a particular number of spokes, chakra being a Sanskrit word that conveys the sense of 'wheel'. Sometimes chakras are referred to as lotus flowers, in which case the spokes become petals. Each represents a power point in the human body, and is associated with various internal organs and systems: for example, the heart chakra is held to govern the entire circulatory system. Each is also connected with a colour and certain beneficial stones (see pages 62–68). Yogis regard the chakras as the connections between the plane of purely mental energy and the physical energy level of our body. They are like a radio set that receives inaudible electromagnetic signals and translates them into sound that our bodies can hear.

The chakras govern the flow of spiritual forces, or *prana*, which circulate along three channels (*nadis*) permeating the physical body. The central channel (*Shushamna*) is associated with the spinal cord, while the other two (*Ida* and *Pingala*) are represented as serpents whose looping coils are entwined around the central column.

According to tradition, in a healthy body *prana* should fill the chakras by rising up into them through the central channel of *Shushamna*. Illness is caused when *prana* flows up into the chakras via the serpents, *Ida* and *Pingala*, therefore throwing the chakras out of balance. Therapists also believe that the channels and chakras may become blocked. A blockage stops *prana* from filling the chakras properly, inhibiting natural behaviour or shutting down the immune system; it can also prevent energy from draining away. This can cause symptoms such as over-excitement, hyperactivity or obsessive behaviour.

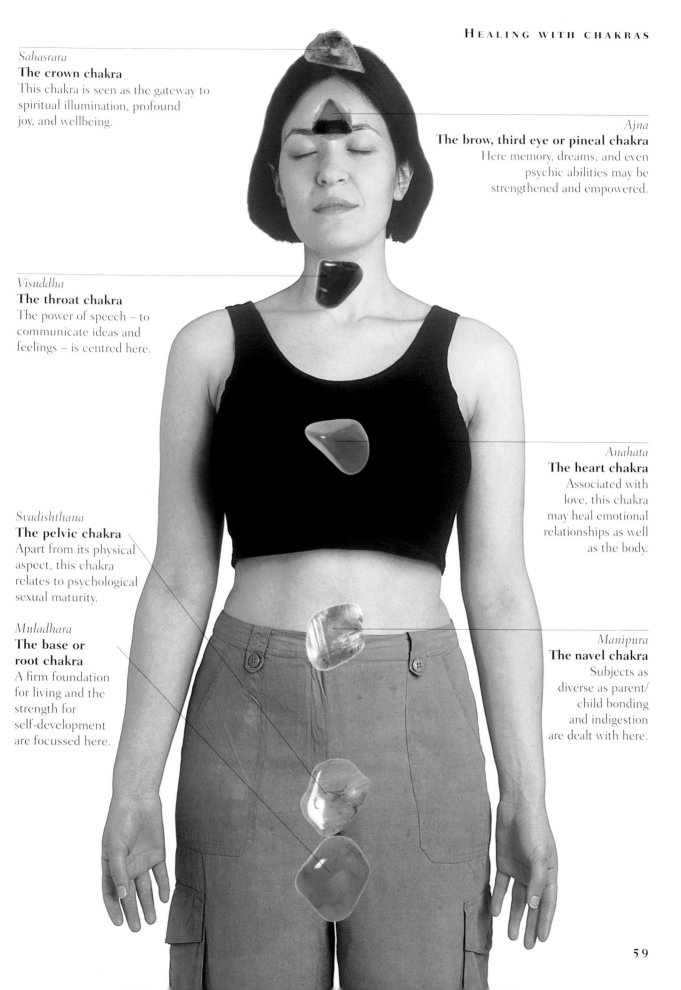

Sahasrara
The crown chakra
This chakra is seen as the gateway to spiritual illumination, profound joy, and wellbeing.

Ajna
The brow, third eye or pineal chakra
Here memory, dreams, and even psychic abilities may be strengthened and empowered.

Visuddha
The throat chakra
The power of speech – to communicate ideas and feelings – is centred here.

Anahata
The heart chakra
Associated with love, this chakra may heal emotional relationships as well as the body.

Svadishthana
The pelvic chakra
Apart from its physical aspect, this chakra relates to psychological sexual maturity.

Muladhara
The base or root chakra
A firm foundation for living and the strength for self-development are focussed here.

Manipura
The navel chakra
Subjects as diverse as parent/child bonding and indigestion are dealt with here.

HOW TO BEGIN

Like so many ancient systems of enlightenment and spiritual healing, the chakra system is itself evolving. The system presented here is based on the millennia-old tradition from India – even so, 100 years of gradual integration with Western ideas has given it several dynamic new slants.

It is worth noting that some books and practitioners have developed this system to the extent that they have altered the positions and powers of the three lower chakras, the *Muladhara*, the *Svadishthana* and the *Manipura*. In this alternative system, the *Muladhara* encompasses not merely the anal region but the genitals as well. In this system, the sex drive is regarded as being fundamental to the human condition. The *Svadishthana* is moved up to the navel and the *Manipura* is given an entirely new position midway between the navel and heart, in the hollow under the ribs, and is known as the solar plexus chakra.

In this ordering, the *Muladhara* combines the powers traditionally split between the two lowest chakras; the *Svadishthana* takes on the powers attributed to the navel (described in the section on *Manipura*) and although the new *Manipura* retains its digestive function, it loses the link with heredity, but gains something of the *Anahata's* (the heart chakra's) ability to balance mood swings.

While we would recommend learning and using the ancient system for home use, you can experiment with the alternative system if that is what a practitioner advises. After all,

When diagnosing chakras, the clearer the crystal you use, the more easily it transmits information.

in the absence of divine revelation, it is only through a process of trial and error that advances are made and traditions improved upon.

DIAGNOSING THE PROBLEM

If you cannot pinpoint the source of an infirmity or if an illness has a complicated variety of causes, you may need help from an experienced practitioner. During a consultation, each chakra is checked in turn, often using a method called 'pendulum dowsing'. This is where the natural pendulum motion of a natural rock crystal, with a good pointed termination, determines in which area of the body the problem lies. Because the wheels of the chakras are seen as revolving, the movement of the pendulum – either with or against the chakra's natural motion – provides a clear indication of the state of its health. If the pendulum moves in tune with the chakra's natural motion, it is healthy, if it moves against the chakra's rhythm, it indicates that there is something wrong.

Traditionally, the sexes have different directions of motion: a woman's *Sahasrara* (crown of head) chakra revolves in an anti-clockwise direction (as viewed looking out from within your body), while that of a man rotates clockwise. The next chakra down (*Ajna*, the brow) rotates in the opposite direction to the one above, so a woman's rotates clockwise, a man's anti-clockwise, and so on. So the lowest *Muladhara* (pelvic) chakra rotates in the same way as the top *Sahasrara* (crown of head) chakra.

If you already know what your condition or problem is, it is easy to use crystals for self-healing. On the following pages the chakras are covered in detail, so that you can tell which ailments are associated with each chakra and which stones and colours to use to relieve them. As a general guide, you (or your patient if you are helping a friend) should be lying down comfortably, preferably flat on a bed or couch.

Make sure that the room is comfortably warm and not too brightly lit so that the atmosphere feels soothing and calm. You can even play some relaxing background music, or alternatively light some incense or burn fragrant oils to help you focus on the healing energy of the stones during your therapy session.

Place the relevant stones (you can use as many or as few as you like) either just on the one chakra where the problem lies, or preferably, on all the chakras leading up to it. Place the stones on the lowest chakra first, and move upwards. Concentrate on the stones' energy, and try to visualize them working in harmony with the *prana* and each chakra, thereby healing the area as the energies converge. The session should last between 15 to 45 minutes, depending on how much time you have.

When you remove the stones at the end of the session, always remove the highest one first, leaving the lowest stone until last. The same technique, as described above, can also be used as a general tonic to strengthen and tone up the entire system – simply by placing a different beneficial stone on each of the seven chakras.

THIS chakra is most often imagined as a lotus flower with white petals, no wider than your head, floating – as if your hair were the surface of a lake of crystal clear water – on the very topmost part of your scalp. The stem of this flower (which is generally visualized as having 1,000 petals) descends in a straight line through the body to the groin, where it is very firmly rooted.

While it is the aspiration of mystics to raise their consciousness from the root to the flower, there to bask in the bright sun of spiritual illumination, we need only remember the picture of the serene lotus blossom to calm our minds. Twirling smooth pieces of stones such as kunzite and charoite, while we imagine the scene where the sacred lotus grows, infuses them with a talismanic power that can help calm our nerves whenever we need it in our day-to-day lives.

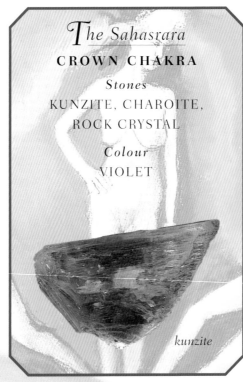

The Sahasrara
CROWN CHAKRA
Stones
KUNZITE, CHAROITE,
ROCK CRYSTAL
Colour
VIOLET

kunzite

Stimulating the *Sahasrara* in this meditative way can be used as a general tonic, and is recognized as having a powerful effect throughout the body, mind and spirit. Although powerful, the *Sahasrara*'s influence can take a while to be noticed because it attempts to harmonize and balance all the chakras beneath it simultaneously.

However, these subtle benefits are cumulative and are generally regarded as the key to maintaining good health. The *Sahasrara* is also regarded as the gateway through which spiritual energy can enter the physical body. If it becomes blocked in some way, problems arise because spiritual forces that normally flow into and through us – and our lives – can no longer invigorate and cleanse us. Such problems often seem to manifest themselves in symptoms such as lethargy and apathy, alienation from other people and loneliness.

Other conditions that are regarded as particularly susceptible to crystal healing using this chakra include stress-related insomnia and learning difficulties caused by mental disorders. People who tend to live in a fantasy world, or those who shy away from involving themselves fully in the process of living, preferring instead to dwell in some fondly imagined utopia, may also find themselves better able to confront the world on its own terms after stimulating this chakra.

CHAKRA HEALING FOR INSOMNIA

To reinforce chakra healing for insomnia,
place the stone beside your bed (or under the pillow).
As you settle down to sleep, picture the thousand petalled lotus
flower that symbolizes this chakra, bathed in warm sunshine and
floating on the gentle swell of a crystal clear lake.

ENERALLY regarded as being situated between the eyebrows in the centre of the forehead and just inside the skull, this chakra has also been identified with the pineal gland.

Sometimes called the Third Eye, the *Ajna* is at the heart of our ability to imagine, dream and see visions. It is an eye in the mind that sees the world as full of hidden meaning and exciting mystery. It is also the centre of psychic and spiritual intuition. This chakra is often stimulated in an attempt to develop psychic abilities such as telepathy, precognition, clairvoyance, empathy, psychometry and even psychokinesis.

Healing this chakra can help to ease the suffering caused by persistent or severe headaches and migraines. Recurring

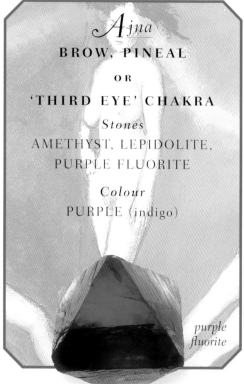

Ajna

BROW, PINEAL

OR

'THIRD EYE' CHAKRA

Stones

AMETHYST, LEPIDOLITE, PURPLE FLUORITE

Colour

PURPLE (indigo)

purple fluorite

nightmares and sleeping disorders such as insomnia, as well as depression and anxiety which can also be treated in this way. The eyes, ears, and nose are also said to benefit from treatments administered here.

Conditions such as earache, sinusitis and blurred vision are all traditionally treated by stimulating the *Ajna*. Even a nasty head cold, with the misery of a blocked nose and runny eyes, can be eased by a gentle massage between the brows with tumble-polished stones such as amethyst or lepidolite. Stimulating the *Ajna* can also help to improve the memory, and is often particularly recommended for students studying for exams, or indeed, for anyone who feels their memory could be improved.

Results cannot be expected overnight of course, but significant effects may begin to be noticeable after two or three sessions.

The combination of amethyst (brow chakra) and aqua aura (throat chakra) promotes both mental insight and the eloquence to share it with other people.

LOCATED in the throat, this chakra is intimately connected both with the power of speech and with the sense of self-worth that gives us the power to express ourselves. Many people who feel shy and unable to share their feelings with others benefit from stimulating the *Visuddha*.

Any blockage in the channel which allows us to voice our opinions and give vent to our emotions can cause deep anxiety and unhappiness, cutting us off from the world in which we live. Because this blockage prevents fluent self-expression, it also prevents other people from getting to know the sufferer, making a spiral of deepening social isolation a real possibility. Working with this chakra can help alleviate these problems by reinforcing self-confidence and also opening up the mind's natural channel of communication – speech.

The *Visuddha* may also be stimulated in order to foster eloquence and the ability to make yourself perfectly clear. Anyone having to make a speech, give a public presentation, or even engage in one of the performing arts that require a strong voice, can benefit from working on the *Visuddha*. Some people actually use this chakra to help tone down a tendency to be arrogant, self-righteous and overbearing.

Speech defects such as stammering may respond particularly well to treatment here.

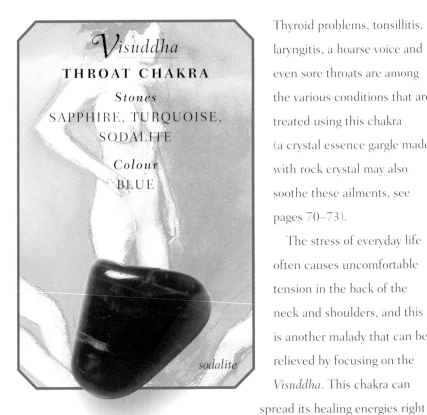

Visuddha
THROAT CHAKRA
Stones
SAPPHIRE, TURQUOISE, SODALITE
Colour
BLUE

sodalite

Thyroid problems, tonsillitis, laryngitis, a hoarse voice and even sore throats are among the various conditions that are treated using this chakra (a crystal essence gargle made with rock crystal may also soothe these ailments, see pages 70–73).

The stress of everyday life often causes uncomfortable tension in the back of the neck and shoulders, and this is another malady that can be relieved by focusing on the *Visuddha*. This chakra can spread its healing energies right down through the shoulders, flowing along the limbs and into the hands. Like the voice, the hands are a particularly important vehicle for self-expression, whether they are used directly in body massage or indirectly in creative art.

All the stress and tension of carrying the weight of the world on your shoulders can be eased by a single crystal.

SITUATED at the heart, the effect of stimulating this chakra can be felt throughout the body, just as the blood pumping through our veins reaches every extremity.

In physical healing, the *Anahata* is widely used to counteract problems with the circulation of blood. For example, low blood pressure can be treated with a light green stone (such as chrysoprase); and high blood pressure can be treated with a dark coloured specimen (such as dioptase) – but as always, ensure conventional doctors are consulted.

Practitioners have found that this chakra can provide a gentle yet comprehensive tonic for anyone recovering from a heart attack or a stroke, and also use it to help patients combat anaemia, angina and hardening of the arteries. Through its association with the

Anahata
HEART CHAKRA
Stones
EMERALD, JADE,
AVENTURINE, MALACHITE

Colour
GREEN

aventurine

thymus gland, the *Anahata* is also stimulated in an effort to boost the immune system and fight against infections.

Healthy individuals can use this chakra to 'fortify' the heart, especially if any unusual physical activity (such as a holiday or sporting weekend) is planned, and there is a need to feel robust and energetic. Similarly, if a period of increased emotional stress is foreseen, then working with the *Anahata* may help to strengthen the emotional balance and prevent any loss of temper or feelings of bewilderment. The hectic run-up to Christmas, for example, is a time when many people would benefit greatly from a little assistance from a resilient heart chakra.

Mood swings (even manic depression) may be particularly responsive to treatment here. It is also thought that women who suffer from pre-menstrual tension can also benefit from the stimulation of the *Anahata*, especially if they commence daily sessions several days before their period.

Of course, the heart is also associated with love, and applying stones here can help us to recognize (and share) our feelings for colleagues, friends and family, as well as partners. The *Anahata* may also help people come to terms with past mistakes, and even find the way to love themselves – which, in itself, can be a fount of self-healing.

With a little practice you can feel your pulse throughout your body, and also sense the flow of the heart chakra's healing energy.

THE name *Manipura* translates as 'city of jewels', which gives an impression of the glories that a clairvoyant might see when contemplating this navel chakra.

The navel chakra is particularly important. While of little practical use once the umbilical cord has been severed, the navel is still a vital part of our being. It is the emblem of the maternal bond, the love that enables a parent to sacrifice itself for the sake of its child. This powerful energy can be tapped through this chakra and it helps us to face up to our responsibilities towards all those who need our help – not only other people but all the animals that share our world.

Anyone wishing to have a more active social life, especially with a view to attracting a partner to settle down with, can use this chakra to boost their personal magnetism. Emotional insecurity and panic attacks are

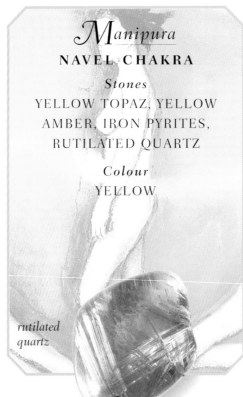

Manipura
NAVEL CHAKRA

Stones
YELLOW TOPAZ, YELLOW AMBER, IRON PYRITES, RUTILATED QUARTZ

Colour
YELLOW

rutilated quartz

conditions that can have a devastating effect, but both respond particularly well to treatment here. Obsessions (especially a compulsion to keep increasingly tight control) and phobias are also treatable here, and very favourable results have been reported.

A wide range of genetic disorders and inherited illnesses are also treated at this chakra point and, although healers tend to stimulate it to ease discomfort rather than effect any particular cure, the effort can still be very beneficial indeed. People interested in exploring past lives can use the navel chakra as a gateway to discovering previous incarnations. However, its power is greatest when contemplating the lives of your own direct ancestors – a fascinating journey which can lead back to the beginnings of time itself.

This chakra is also responsible for digestion and healers stimulate it to treat conditions such as ulcers, irritable bowel syndrome, gastritis, morning sickness, indigestion and trapped wind. Anorexia and bulimia are among a range of conditions that may respond well to working with this chakra. Diabetes, diseases of the liver such as hepatitis, and also gall bladder problems such as gall stones all come under the influence of the navel chakra. It is usually stimulated as a complementary therapy to conventional medical care.

You can place the stone in the navel to promote maternal bonding energy, or place it beside the navel for treating the digestive system.

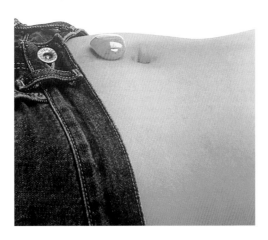

I DENTIFIED with the primary sexual organs, this chakra is associated both with the physical process of conception and with the powerful psychological urges which are a hallmark of sexual maturity in humans.

The instinct for reproduction is particularly strong in human beings and many people experience some difficulty in coping with the pressure this exerts on our behaviour. Whether they find that their sex drive is inhibited or overactive, the *Svadishthana* is employed as a counterbalance, and working with this chakra can help to bring these urges into a healthy and naturally harmonious state.

Some modern practitioners regard the sexual chakra as being synonymous with the base chakra, but this is merely due to the Western preoccupation with the idea of sex being the root of all adult behaviour. Sex was always recognized as a powerful force by the yogis, who saw in it a sublime metaphor for an ecstatic union with the divine.

Although the mood swings that characterize adolescence are best treated in the *Anahata*, stones appropriate to the *Svadishthana* – tiger's eye in particular – can be beneficial to developing teenagers (even if only carried, for example, in the front pocket of a pair of trousers). The same holds true for people entering the menopause.

Svadishthana
PELVIC CHAKRA
Stones
ORANGE JASPER, TIGER'S EYE, SUNSTONE

Colour
ORANGE

orange jasper

Adult women who experience acute period pains are also offered treatment at this chakra point, a series of sessions being held in the days prior to the period itself. And, of course, an expectant mother can use this chakra to strengthen herself against premature birth and labour pains. Although wary of making extravagant claims, practitioners do attempt to treat infertility – in both women and men – by stimulating this chakra. By using stones to bring fresh energy into this area, healers seek to assist in conception where fertility is low. The stones can also help to alleviate impotence.

The *Svadishthana* is also activated to speed up and strengthen conventional medical treatment for prostate conditions. Other ailments of the urogenital system, ranging from venereal disease to cystitis and kidney problems can also be treated here.

HELPING PREMENSTRUAL TENSION

To reinforce chakra healing for premenstrual tension, carry one of the stones listed above with you. When you have the opportunity, gently rub the stone over your lower abdomen, starting just below the navel, move in a large circle toward your left, and then let the circle become a spiral towards the middle. Focus on the warmth flowing from the stone, bringing deep comfort and relaxation.

THE name *Muladhara* is composed of two words meaning 'root' and 'support' respectively. This name reminds us of the image of the glorious white lotus flower that blossoms in the *Sahasrara* chakra (at the top of the head) – here, in the *Muladhara*, can be found the strong, hungry roots that delve deeply into the thick, fertile silt at the bottom of the lake and hold the whole plant, with its long stem, securely fixed in position.

The base chakra represents a firm foundation for our entire

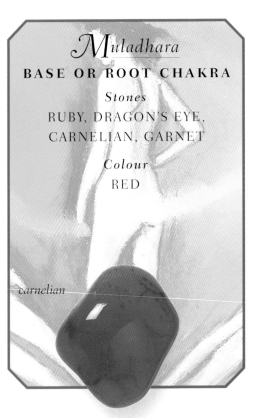

Muladhara
BASE OR ROOT CHAKRA

Stones
RUBY, DRAGON'S EYE,
CARNELIAN, GARNET

Colour
RED

carnelian

being: its primal energy is what keeps us going when we are alone and things get really tough.

This chakra is situated at the base of the spine and is associated with the anus. Crystals should be placed so that they are actually resting on the ground between the legs, their position corresponding to the base of the spine. The colon also comes under its sphere of influence and medical conditions affecting this part of the body are suitable for therapy using the *Muladhara*. Ailments such as constipation,

A CHAKRA MEDITATION

Anyone who can raise their *prana* up through the central channel (*Shushamna*) to the crown chakra achieves the ultimate goal of spirituality. To attain this, a profound system of meditation is necessary. While you may not achieve this highest level of spirituality quickly, by undertaking the following series of seven meditation sessions you can certainly be assured of a rewarding experience. You can choose any of the beneficial crystals associated with each chakra. Your first session should last about 15 minutes, building up to longer sessions of 45 minutes as the week progresses. Always spend 15 minutes focusing on the new chakra for that day.

For the first session, place a single stone on the lowest base chakra. Contemplate this chakra's role in your personal life and in the world at large.

For the second session, position one stone on the lowest chakra and another on the chakra above it (the pelvic chakra). Begin your contemplation with the lowest chakra and when comfortably complete there, imagine the energy of *prana* flowing up the central channel and into the next chakra. During this session, think about the role of these chakras as well as trying to be sensitive to how you feel about them, as in this way your intuition may help you to find enlightenment.

As you remove the stones, (in reverse order, taking off the highest one first), it is also important to imagine the *prana* lowering through the central channel. This completes the exercise by reintegrating you with the world at large yet also ensures that all the chakras that have been stimulated are cleansed, strengthened and ready for use.

Work upwards through the chakras on each day. On the seventh day, all the chakras are covered and the *prana* is symbolically raised, step-by-step, to the *Sahasrara*, to shine as the jewel in the highest lotus. Although the *prana* may fall as the stones are removed, you should find it easier to raise it again at will.

haemorrhoids and diarrhoea are treated here, as are paralysis and other spinal injuries such as slipped discs and backache.

Problems with the legs and feet also benefit from the healing energy radiating from this chakra and make it especially popular with healers, who regularly use it to ease the suffering caused by conditions such as sciatica, rheumatism, arthritis, varicose veins and bunions. Sporting injuries that affect the legs and feet (such as torn tendons, pulled muscles and sprained ankles) are also

among the conditions which are particularly susceptible to the therapeutic energies that are released by this chakra.

Furthermore, these benefits are often extended to assist with faster healing and strengthening of broken bones in the legs and feet. Because of its effect on the lower limbs, athletes and sports players can also use the base chakra to help build up and tone this part of their body so that they obtain the maximum benefit from training and can compete more successfully.

Crystal essence remedies

Also termed gem elixir therapy, this is a modern version of an ancient tradition in which special stones are immersed in pure water to extract their healing essence in much the same way that herbs are steeped to make a herbal tea.

The theory behind crystal essence remedies lies in the belief that the stones' curative properties percolate into the water and are transferred to the liquid, which absorbs and stores these subtle forces. The liquid is said to take an impression of the energy of the stone, and the resulting fortified liquid can be drunk to combat a specific ailment or simply taken as a preventative and general tonic. The illnesses that respond best to this therapy are those thought to be caused by a mental, emotional or spiritual blockage or imbalance. This form of treatment is particularly good for stress-related conditions.

Some remedies, such as herbal or homeopathic medicines, can produce powerful reactions in the patient. These are akin to side-effects and are seen as evidence of the patient's body responding to the medicine and changing to a healthier state. Crystal essence remedies will not cause people any obvious side effects, and are seen as being comparable to flower essence remedies.

Crystal essences, particularly rock crystal, can also be used to mist the home and house plants as well as being poured into bath water to allow you to bathe in the healing energy of the stones. You could even make up an essence as a gift for a relative or friend who is unwell or under some kind of stress – it's a wonderful way of showing that you care!

Before preparing your crystal essence remedy, refer to the comprehensive list of common ailments and conditions on pages 74–75 to find out which stone you need to help your condition, be it emotional, mental, physical or spiritual.

HOW TO MAKE YOUR OWN ESSENCES

First, choose a stone or stones that are hard and smooth. This is important because the stones and the glass container should be hygienically clean, so flaky or powdery stones, or those that have cavities which could harbour micro-organisms, are not suitable for making essences. The stones must also be pure: that is not embedded in, or alongside any other potentially dangerous mineral in the same specimen. And remember, where more than one form of a particular mineral exists, it is only the most solid or crystalline form which should be used.

Once selected, some practitioners like to purify, balance and charge their stones before

Apophyllite

using them to make an elixir. You can do this by placing them in the midst of a rock crystal cluster for up to two hours. This process can also be repeated after the stones have been taken out of the liquid, cleaned and dried. To prepare your crystal essence:

1. Place the appropriate stone in a glass tumbler, jar or bottle.

2. Fill the vessel with enough water (preferably bottled mineral water) to completely cover the stone.

3. Cover the container to prevent contamination. Before use, the liquid should be kept as free as possible from contamination by micro-organisms etc.

4. Put the container in a safe place where it can bask in as much direct sunlight as possible, or in moonlight if preparing an essence that is meant to calm, relax or boost receptivity. You can leave it to steep for anything from one minute to 12 hours. (For example, if you prepare a moonlight essence before going to bed, you can take a few drops in the morning). The longer the immersion, the stronger the essence will be.

5. Strain the liquid from the stone.

USING THE CRYSTAL ESSENCE REMEDIES

Before you drink the essence (and afterwards) a few minutes' rest is recommended as, although the elixir helps the healing to happen, it is your own life-force that actually brings about the healing.

If the stone has only been soaking for a minute or two, you may drink all the liquid. If it has been submerged for up to one hour, take a few sips only; if submerged for up to six hours, take one small sip only. If the stone has been immersed for over six hours, take a few drops of the liquid only.

You can also mix the elixir with alcohol to form a tincture, which will extend its shelf-life. In this case, you should use distilled water instead of bottled water and, after draining the liquid from the stone, add an equal quantity of brandy. Take only a few drops of the tincture in one dose.

You can repeat the dose several times a day but, of course, if you still feel run down then qualified advice should be sought. It is usually advisable to use the essence the same day unless you bottle it and store it in the fridge,

Preparing a refreshing crystal essence tonic can take only as long as it takes to make a cup of tea.

CRYSTAL ESSENCES FOR PETS

Although exotic or delicate creatures need professional treatment, robust animals such as dogs and cats can benefit from home therapy with essences.

Doses can be administered in their drinking water, with quantities ranging from half-human measures for large dogs, to one-tenth for small dogs and cats. Milky quartz, for example, can help an animal settle down in a new home, especially if it has come from an unhappy situation. Use amethyst crystals to integrate a new animal into an existing menagerie. Citrine essence is a good all-round tonic and can be given weekly; a listless animal may benefit from a short dose of carnelian while an over-energetic or disobedient animal can be given smoky quartz. Do consult a vet though, if symptoms persist or are worrying in any way.

Cats in particular have minds of their own, and kittens are especially fond of clawing the furniture. One way to exercise a measure of control without leaping from your chair every two minutes is to use a water pistol (note: choose one from the weakest end of the range, as nobody should aim a powerful water jet at any animal).

The cat's passion for cleanliness will soon persuade it that getting soaked fur every time it scratches is too high a price to pay (and remember, custom-made scratching posts are sold in pet shops). You can also use rock crystal essence to help control your pet so that it eventually gets the message.

where it will last for several days. Tradition dictates that it should never be frozen into ice, the crystal form of water, because this destroys its curative properties. If more treatments are needed, prepare fresh essence – just clean and dry the stone before re-using.

Safety guidelines
Many minerals will react with water on a chemical level. (Common salt, for example, dissolves in water). Because some minerals, such as galena, are poisonous if taken internally, some stones mentioned in this book are NOT appropriate for crystal essence remedies. So if in doubt, follow our recommended suggestions overleaf or seek professional advice.

Stones to relieve common ailments and conditions
Whether your pain is physical, mental, emotional or spiritual, crystal remedies can help. Although you should always see your doctor if you have any health problems, especially if an illness is prolonged or suddenly severe, crystals can help to relieve many minor ailments and to promote general wellbeing. Now that you know how to prepare a crystal essence remedy, you can refer to the comprehensive list of common problems overleaf to choose your crystal cures. In addition to making the essences, to treat physical pain you can also try rubbing the crystal against the part of your body that is causing distress.

Problem-solvers

Physical

PROBLEM	ASSOCIATED STONE
Headache, neuralgia, sunburn, high temperature, minor burns and scalds, muscular aches, sprains	red carnelian red jasper
Tense neck and shoulders, sore throat, laryngitis, halitosis	malachite aventurine
Warts, dry and chapped skin, colds, flu	tumble-polished rock crystal
Indigestion, swollen breasts during menstruation	moonstone milky quartz
Backache, poor circulation	citrine rutilated quartz
Menstrual cramps, irritable bowel syndrome	rock crystal egg
Sciatica, lower back pain, numbness from sitting for too long at a desk, incontinence	chrysoprase peridot
Premature ejaculation, vaginal dryness, low libido	snowflake obsidian moss agate
Rheumatism, lowered resistance to disease, overeating	sodalite turquoise
Toothache, acne, bone fractures, joint pain	hematite obsidian
Cramp, broken veins, hyperventilation, phobias, panic attacks	amethyst lepidolite
Corns, verrucae, chilblains, water retention, dependency on caffeine, alcohol or tobacco	chrysocolla aquamarine
Sluggishness, lethargy, to invigorate the entire body	rock crystal
To promote healing	watermelon tourmaline
Inability to relax	malachite

Emotional/mental

PROBLEM	ASSOCIATED STONE
Difficulty in overcoming sexual inhibitions	garnet
Insomnia, stress	rhodochrosite
Lack of confidence	aventurine
Overactivity	star sapphire
Nervous tension	apatite
Fear of the unknown	tourmalinated quartz
Sexual anxieties	smoky quartz
Insecurity	nephrite jade
Depression	rutilated quartz
Lacking understanding	blue lace agate
Poor judgement	sodalite
Poor memory	amber
Fear of making mistakes	azurite
Nerves, anxiety	epidote, diamond
Inability to let go of attachments	tanzanite
Repressed anger and fear	rose quartz
Lack of inspiration or receptivity	moonstone
Negative self-image	citrine
Difficulty in expressing love or emotion	dioptase
Difficulty in raising consciousness	bloodstone

Spiritual

PROBLEM	ASSOCIATED STONE
Inability to transcend dogma	charoite
Unable to feel universal love	iolite
Lack of spiritual insight	labradorite
Problems in developing psychic abilities	sapphire
Inability to recognize spiritual messages	lapis lazuli

STAR CRYSTALS

Crystals have long been linked with astrology. Just as the planets rule various signs of the zodiac, there are gems that are governed by both the planets and the star signs. The stones allocated to the zodiac signs here are actually closer to the traditions of ancient astrology than the 'traditional' birth stones, which are primarily an invention of merchants and jewellers. This section of the book tells you all you need to know about your own birth stones and how to use them in your everyday life. There are myriad ways in which crystals can help you to bring out the best facets of your personality and to successfully deal with just about anything that life throws at you! For more specific problems connected with your star sign, turn to pages 128–134, for advice.

Which are your star crystals?

Depending on your star sign, you will have three major crystals that are appropriate for you: a ruling crystal (like a ruling planet), a sun sign crystal (like a birth stone) and a moon sign crystal. It is easy to work out your sun sign, as it is traditionally your zodiac sign, which relates to the date that you were born.

ARIES:	**LEO:**	**SAGITTARIUS:**
March 20 – April 19	July 23 – August 22	November 22 – December 21
TAURUS:	**VIRGO:**	**CAPRICORN:**
April 20 – May 20	August 23 – September 22	December 22 – January 19
GEMINI:	**LIBRA:**	**AQUARIUS:**
May 21 – June 20	September 23 – October 22	January 20 – February 17
CANCER:	**SCORPIO:**	**PISCES:**
June 21 – July 22	October 23 – November 21	February 18 – March 19

YOUR MOON FINDER CHART

YEAR					Jan	Feb	March	April	May	June	July	Aug	Sept	Oct	Nov	Dec
1920	1939	1958	1977	1996	Tau	Can	Can	Vir	Lib	Sag	Cap	Aqu	Ari	Tau	Can	Leo
1921	1940	1959	1978	1997	Lib	Sco	Sag	Cap	Aqu	Ari	Tau	Can	Leo	Vir	Sco	Sag
1922	1941	1960	1979	1998	Aqu	Ari	Ari	Gem	Can	Leo	Vir	Sco	Cap	Aqu	Ari	Tau
1923	1942	1961	1980	1999	Gem	Leo	Leo	Lib	Sco	Cap	Aqu	Ari	Tau	Gem	Leo	Vir
1924	1943	1962	1981		Sco	Sag	Cap	Aqu	Ari	Tau	Gem	Leo	Lib	Sco	Sag	Cap
1925	1944	1963	1982		Pis	Tau	Tau	Can	Leo	Lib	Sco	Sag	Aqu	Pis	Tau	Gem
1926	1945	1964	1983		Leo	Vir	Lib	Sco	Sag	Aqu	Pis	Tau	Can	Leo	Vir	Lib
1927	1946	1965	1984		Sag	Cap	Aqu	Pis	Tau	Gem	Leo	Vir	Sco	Sag	Aqu	Pis
1928	1947	1966	1985		Ari	Gem	Gem	Leo	Vir	Sco	Sag	Aqu	Pis	Ari	Gem	Can
1929	1948	1967	1986		Vir	Sco	Sco	Cap	Aqu	Pis	Tau	Gem	Leo	Vir	Lib	Sag
1930	1949	1968	1987		Cap	Pis	Pis	Tau	Gem	Leo	Vir	Sco	Sag	Cap	Pis	Ari
1931	1950	1969	1988		Tau	Can	Can	Vir	Lib	Sag	Cap	Pis	Ari	Gem	Can	Leo
1932	1951	1970	1989		Lib	Sag	Sag	Aqu	Pis	Tau	Gem	Can	Vir	Lib	Sag	Cap
1933	1952	1971	1990		Pis	Ari	Tau	Gem	Can	Vir	Lib	Sag	Cap	Aqu	Ari	Tau
1934	1953	1972	1991		Can	Vir	Vir	Lib	Sag	Cap	Pis	Ari	Gem	Can	Vir	Lib
1935	1954	1973	1992		Sco	Cap	Cap	Pis	Ari	Gem	Can	Vir	Sco	Sag	Cap	Aqu
1936	1955	1974	1993		Ari	Tau	Gem	Leo	Vir	Lib	Sco	Cap	Pis	Ari	Tau	Can
1937	1956	1975	1994		Leo	Lib	Lib	Sag	Cap	Pis	Ari	Tau	Can	Leo	Lib	Sco
1938	1957	1976	1995		Cap	Aqu	Pis	Ari	Tau	Can	Leo	Lib	Sco	Cap	Aqu	Ari

DATE OF BIRTH	1	2	3	4	5	6	7	8	9	10	11	12	13	14	15	16	17	18	19	20	21	22	23	24	25	26	27	28	29	30	31
	0	1	1	1	2	2	2	3	3	4	4	5	5	5	6	6	7	7	8	8	9	9	10	10	10	11	11	12	12	1	2

CALCULATING YOUR MOON SIGN

Using the table on the left, look for the year of your birth, read across to the month column and note the star sign listed. Take the number of the day you were born and note its corresponding number beneath. For example, if you were born on the 12th, your number is 5. To find your moon sign simply count 5 signs on (using the list of sun signs directly opposite, and reading from top to bottom) from the sign according to the date of your birth – for example if this is Sagittarius, your moon sign is Taurus.

HOW TO USE YOUR BIRTH STONES

In this section, we have used the terms 'promotes' and 'balances' to describe how your birth stones can work to accentuate the most positive characteristics imbued by a star sign and to lessen the negative traits attributed to its influence. Leos, for example, may use imperial topaz to bring out their leadership qualities. They can also use it to subdue their egotistical tendencies!

If you cannot find any of the stones suggested, you can always use a simple rock crystal wand instead – many people find that it performs almost as well as the real thing.

Sleeping with crystals beneath the pillow or beside the bed, is a good way of contacting deeper levels within yourself. On awakening, they can also provide a link with your dreams.

Handling crystals, like worry eggs, is a way to dispel nagging fears or gather energy. Wearing them as jewellery is a time-honoured way of bringing them, literally, close to you. Reading about the stones, looking at them, handling them and having them around you are all conducive ways to experiencing their unique properties and discovering how useful they can be in your everyday life.

If any of the stones for your star sign are not commonly used in jewellery then you can buy a ready made silver wire spiral, which comes in a variety of sizes, and place the crystal inside it.

The choice of silver for the metal is particularly appropriate because this is the metal of the moon, which will boost your receptiveness to the influence of the stone you are wearing. Similar constructions made from gold and other metals can be more difficult to come by and may 'muffle' the stone's energy, rather than amplifying its virtues.

If body-piercing appeals to you, try obtaining beads of the appropriate stones with which to close the ends of the body jewellery – having first ensured that such stones are safe to wear!

If you were born within a few days of the start or end of your zodiac sign (technically termed the cusp), it is best to have your sun sign calculated carefully before you buy expensive birthstone jewellery. The dates quoted by magazines, jewellers and the like are really only rough guides.

Aries

Key word

PASSION

Characteristics

ENTHUSIASTIC,
HOT-TEMPERED, IMPULSIVE

Ruling planet

MARS

Ruling stone

RUBY

Sun stone

IMPERIAL TOPAZ

Moon stone

WHITE MARBLE

Ruby

ARIES (the ram) is ruled by the planet Mars, which represents competition and war. Although originally a protective god of nature and guardian of the flocks, the Roman god Mars evolved into a war-god and came to be seen as hot-headed and volatile. Mars is pioneering, self-assertive, competitive and dynamic, and governs battles of all kinds, such as on the sports field or, figuratively, in the boardroom or even in the bedroom!

Aries is the sign of new beginnings, associated with the thrusting new growth of spring. At its best, Aries symbolizes leadership qualities, outspoken frankness and a positive attitude to life. On the down side, those born under this sign can be aggressive, prone to bullying those weaker than themselves and easily bored.

Ruling stone
RUBY

While red stones in general are emblematic of Mars, ruby is its primary stone. This bright, precious gem is a beautiful embodiment of nature – red in tooth and claw – and it can be used in any situation involving competition or business, physical strength or endurance, men in general, or where Arien qualities are needed. Ruby can be used both to promote the positive traits of Aries and to minimize or balance the less desirable ones.

Ruby is a red variety of corundum (the blue variety is sapphire), which is usually found as a six-sided prism. Its name is derived from the Latin word *ruber* meaning red. Although gem

quality crystals are highly valued, less expensive stones are perfect as birthstones.

Throughout the ages ruby has been hailed as the king of stones, not only because of its rarity but also for its red colouring, a warning of the bloody road to conquest and dominion. Ruby is also associated with healing wounds, as well as with promoting youth and vitality. A 13th-century king of Siam reputedly died at the age of 90 with the complexion of a youth, a remarkable achievement attributed to the fact that he rubbed his face and neck with a ruby each morning and night!

Rubies are best worn as jewellery – particularly as earrings (even nosestuds!) since Aries is associated with the head. Ideally, wear the stones in any competitive situation or when stamina, determination and forcefulness are required.

Sun stone
IMPERIAL TOPAZ

Imperial topaz is a supremely positive stone that is especially symbolic of an enterprising spirit and vitality. Those with Aries as their sun sign are often born leaders: they are honest, have boundless energy, enthusiasm and the ability to inspire others. Imperial topaz promotes these qualities and also helps to bring their negative

Wear rubies in any competitive situation or when stamina, determination and forcefulness are required

sides – recklessness and egotism – under control. Imperial topaz has been prized for its ability to bring the favours of influential people to the wearer, which is notable because Aries' loyalty is the purest in the zodiac. Aries people are loyal by nature, not because they have any ulterior motive.

The Sanskrit word for topaz means 'fire' and the stone is symbolic of this element. According to ancient tradition, topaz could diminish heat as well as provide it. So powerful was this property that specimens were thought to suppress the libido – indeed topaz is still regularly used to massage the forehead in order to calm the mind and cool the emotions. Another allied medieval belief was that topaz could cure fevers (this was perhaps, a remarkable precursor to the homeopathic methodology of fighting fire with fire), a feature that is particularly appropriate since Aries people are especially prone to fever, headaches and migraines.

Ariens, usually too busy thinking about tomorrow's ventures to rest properly, also tend to suffer from insomnia or nightmares. For a sound night's sleep, place topaz under the pillow – its reputation for warming the darkest of nights makes it a popular remedy for these problems. Imperial topaz is one of the golden or yellow gemstones allocated to the *Manipura* (navel) chakra (see page 66), which again corresponds to the element of fire. It also relates to the digestive system and the correct absorption of energy. Ariens have the highest energy requirement of any of the signs but they often fail to take the time to eat sensibly. The calming effect of topaz can help to slow them down a little and encourage healthier eating habits.

Topaz is best worn as jewellery, although keeping a few small crystals close by while at work or during the night may also prove effective. The stone's warm, gentle brilliance urges Ariens to direct their joy in life outwards and to appreciate and cherish other individuals. Although topaz crystals are found in a wide range of colours, from blue to green to pink, the distinctive golden-coloured, transparent stones are the most common.

Moon stone

WHITE MARBLE

Those born with the moon in Aries can often be restless and impetuous. Their moods and feelings may swing dramatically and the way that they can change from having one point of view to believing the complete opposite can be either invigorating or positively

Ariens tend to suffer from insomnia or nightmares. For a sound night's sleep place topaz under the pillow

A protective stone, white marble is helpful for those under emotional stress

frightening! The solidity and hardness of white marble acts to counterbalance the flighty, changeable natures of moon-influenced Ariens. As it is a very protective stone, it is also very useful for anyone who frequently finds themselves under emotional stress.

White marble contains a large proportion of calcite, with assorted other minerals producing its different colours and distinctive appearance. It has been used for millennia in the construction of sacred and noble buildings and especially in sculpted statues.

Although it is unlikely – these days, anyway – that the vast majority of people would want, let alone be able to afford, marble floors or columns for their houses, marble is readily available in far less ostentatious and expensive forms. A worry egg would be the ideal way to use the stone, while a small but tasteful sculpture or even a paperweight or bookend would serve almost as well.

Of course, if you have the cash and inclination for something larger and more flamboyant, a marble front step to your home might prove beneficial, certainly for the time you are inside. Marble chopping blocks and mortars and pestles for the kitchen are also easy to find. There are other ways of incorporating marble into your home and life: try scouring market stalls and antique shops – you may unearth a few surprising bargains!

Taurus

Key word
STRENGTH

Characteristics
**PATIENT,
CONSERVATIVE, PLACID**

Ruling planet
VENUS

Ruling stone
EMERALD

Sun stone
AMBER

Moon stone
DESERT ROSE

Emerald

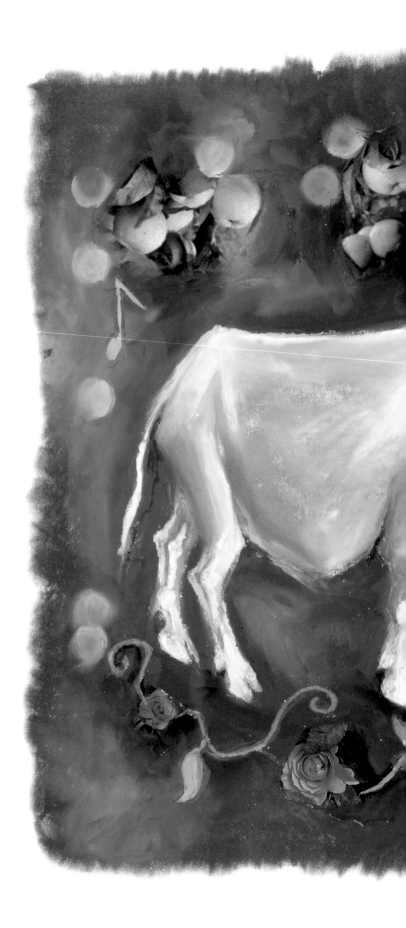

TAURUS (the bull) is ruled by the planet Venus, which represents love and fertility. Venus was a fertility goddess akin to the Great Mother, who ruled the entire natural world. Venus represents the feminine creative principle and maternal, affectionate and nurturing qualities.

Taurus (the bull) is the sign of stability. It encompasses everything that is steadfast, strong and patient. At its best, Taurus symbolizes loyal affection, generosity and sociability, while its worst connotations are emotional dullness and romantic apathy. Taureans are sensual, tactile and home-loving.

Ruling stone
EMERALD

Green stones are emblematic of Venus and emerald is this planet's foremost stone. Emerald has long been held to protect pregnant women, probably due to its association with Venus or Aphrodite. Yet it was also believed to preserve women's chastity and ancient Hindu writings assert its power as an anaphrodisiac – literally a turn-off!

This gem is said to help sustain constancy between lovers and, when kept clean, to attract money and multiply property. It can be used in any situation involving romance, love and women and will help to accentuate the positive traits and minimize the negative ones that are characteristic of Taureans.

Emerald is a transparent variety of beryl, its green colour being due to traces of chromium. It derives its name from a Persian word meaning green and the best stones rival diamonds in value and beauty. History tells us that emeralds were mined in Egypt more than 3,500 years ago. Mines in Columbia, long before the Spanish conquest, once supplied the Inca rulers with these stones, which were revered as representing the green, fertile earth.

In medieval times it was believed that emerald granted the gift of prophesy when placed under the tongue. Indeed, it was said that even an incorrigible liar, confronted with this stone, would have to confess the truth. Both these ideas correspond with Taurus's association with the throat and the organs of speech.

The connection between emeralds and justice and truth is illustrated by a Hebrew writing which asserted that a serpent, then a symbol of deception, could be blinded simply by looking at one. Other notions were that emeralds promoted eloquence, improved the memory and steadied the nerves. Today, emerald is extremely useful in situations where calm, endurance and good memory skills are needed, such as in exams or tests.

Emeralds are best worn as jewellery, around or near the neck. Necklaces, pendants and pendant earrings are all appropriate.

> *Emerald is extremely useful in situations where calm, endurance and good memory skills are needed, such as in exams or tests*

Sun stone

AMBER

Those born under the sun sign of Taurus have an overwhelming need for physical security and this manifests itself in the desire for financial status, such as the ownership of property and solid, good quality objects. Such accoutrements make for a contented bull. Their qualities are stability, patience and conservatism but they can be stubborn, lacking in imagination and have a tendency to be stolid – if not boring!

The sun stone for Taureans is amber. Amber's colour varies through orange to red, the most highly prized specimens being transparent and resembling solid sunlight. Nowadays, the main source of amber in Europe is the Baltic coast. It has been used throughout history all over the world, such as during the Bronze Age, when amber pendants were bound in gold and large pieces were carved into cups. According to Norse mythology, amber was formed by the tears of Freya, goddess of love and fertility, which solidified when they fell into the sea. Homer provides an early written record of amber jewellery when he describes a suitor's gift to the faithful Penelope as 'a golden chain strung with amber beads that gleamed like the sun'. The ancient Chinese perceived amber as the soul of a tiger, left behind after its death.

In fact, amber is fossilized tree resin, usually from conifers about 50 million years old. It feels warm to the touch, is amazingly lightweight and often contains insects or seeds (occasionally even small lizards).

When rubbed, for example with wool, amber gains an electrical charge and will attract small fibres and dust. Its ancient Greek name was *elektron*, from which we derive the word electricity. It is said that dreaming of amber heralds an unexpected windfall or swift reimbursement of loss.

Its great age and organic origin make amber the perfect talisman for Taureans, especially when worn around the neck, since this is the part of the body associated with Taurus. For this reason, amber can help relieve a sore throat. Amber is also reputed to attract material goods – if you wear it to a car boot sale or antique market, it may help you unearth treasures!

Amber is reputed to attract material goods – if you wear it to a car boot sale or antique market, it may help you unearth treasures!

Moon stone
DESERT ROSE

Those born with the moon in Taurus have extreme emotional depth. They may be slow to make decisions or come to definite conclusions but once their minds are made up, these Taureans will pursue their objectives unswervingly. The moon's influence imbues them with sensuality, endurance and a sense of history, but it may also manifest itself in emotional rigidity and a fear of losing control.

Their fundamental stone is desert rose. Formed from minerals left behind when ancient lakes evaporated, desert rose's power is linked to the land masses of the world. As such they serve to steady Taureans' intensity of feeling, helping to 'ground' them and make them more placid.

Moon-influenced Taureans are not averse to using their natural animal magnetism to attract partners and satisfy a high sex drive. People who seek to emulate these bullish traits often use desert rose stones as talismans – however those not born with the moon in Taurus should beware of desert rose, without the Taurean traits of absolute openness and honesty, it may lead them towards extravagance and recklessness.

Desert rose has a reputation for acting in at least two different ways: it can gently invigorate and enliven a sluggish body, mind or spirit, and can aid an understanding of the necessity of change in order to evolve. The stone owes its name to its resemblance to the rose flower – its clusters of rounded crystal plates look very much like petals.

Desert rose is very fragile and therefore it is best not to handle it too frequently. Neither is it wise to immerse it in water as its mysterious desert nature is regarded as being weakened by immersion. The most effective way to use the stone is simply to have it nearby, so that you can look at, admire and occasionally touch it.

Desert rose can gently enliven and invigorate a sluggish body, mind or spirit

Gemini

Key word
PERCEPTION

Characteristics
**INQUISITIVE,
INCONSISTENT, ADAPTABLE**

Ruling planet
MERCURY

Ruling stone
**DOUBLE TERMINATOR
ROCK CRYSTAL**

Sun stone
CITRINE

Moon stone
APOPHYLLITE

Rock crystal

GEMINI'S (the twins) ruling planet is Mercury, the fastest-moving of the planets and the closest to the sun. Mercury, or Hermes, was the messenger of the gods, possessing winged sandals and a winged cap so that he could move with the swiftness of thought from heaven to earth with messages from the gods to men. However, it is important to remember that because of his fleetness, Mercury was also the god of thieves and swindlers!

Gemini is the sign of communication. It embodies the principle of choice – the difference between good and evil, light and dark and so on. Gemini's positive qualities are versatility and an enquiring mind. But Geminis can also be impatient, superficial and prone to trivializing emotion.

Ruling stone
ROCK CRYSTAL

Rock crystal – in all its forms – is emblematic of Mercury and the double terminator rock crystal is the number one stone for Geminis. Fascinating and transparent, it symbolizes clear-mindedness and lucid communication.

The double terminator is effective in any situation that involves clear thinking, lucid communication, siblings and children in general, or where Geminian qualities are needed. It can be used to promote the positive traits and to minimize or balance the less desirable qualities displayed by those born under this sign. In a double terminator crystal, the six sides of the shaft of the crystal slope steeply together to form a point at both ends. Because natural double terminators grow in a semi-fluid environment (such as clay) they are free to develop their characteristic pointed terminations at both ends. Occasionally, rock-based crystals can be snapped free by geological forces and, if conditions are right, continue to grow at both ends and therefore heal the broken end. Some quartz crystals are artificially cut, ground and polished to have a point at each end, but these are not natural and should be thought of as faceted rather than legitimately double terminated.

Double terminator crystals symbolize the uniting of opposites in one body, making them particularly appropriate for Gemini since this sign, more than any other, characterizes duality. The stone also represents two-way communication, between groups, instructor and pupils, individuals, between the mind, body and spirit, and even between the conscious, unconscious and the subconscious.

In stressful situations the double terminator can act like a safety-valve, deflecting stress away from the sufferer. However, the stone does not have a calming effect; rather it acts to strengthen the system so that the individual can continue under high pressure (it is a typical Geminian trait to think of, and try to do, everything at once). As such, it should not be used continuously for this purpose, as it may only serve to disguise the symptoms of

The double terminator can act like a safety-valve, deflecting stress away from the sufferer

genuine nervous exhaustion. The best way to use this crystal is to hold and handle it frequently or to wear it in the form of a pendant or earrings. In the latter case, just make sure that both points are 'free', since with all natural crystals, capping the end may 'cramp' its power.

Sun stone
CITRINE

Those born under the sun sign of Gemini are adaptable, alert, inquisitive and adept at the communication of ideas. Geminis talk and listen to others out of a genuine interest in their thoughts, opinions and ideas. However, they can be very reticent in talking about themselves, although their considerable charm usually prevents this causing irritation.

Of course, there is also the other type of Gemini, the inconsistent busybody who wants to know everyone's business and is quite happy to talk about them behind their backs. Such a Gemini is at best shallow and two-faced, and at worst mentally unstable. While not necessarily dangerous, this Gemini's careless speech can cause a great deal of hurt and upset.

The sun stone for Gemini is citrine, which acts to balance impatience and restlessness. This is a transparent yellow or golden variety of quartz: if possible, try to choose one with a rich golden cast. Heating amethyst or smoky quartz can produce a false citrine that often displays shades of red, orange or bronze: while to some extent this

When used as a meditation stone, citrine helps to clarify thoughts and ideas

reflects the versatile nature of Gemini, the genuine article may be more useful in serving to remind Geminis to be more circumspect and honest in their dealings with others.

Traditionally, citrine protects the wearer from nightmares – an attribute that may be linked to its traditional efficacy in aiding digestion, both physical and mental. It is also reputed to promote psychic awareness and, when used as a meditation stone, helps to clarify thoughts and ideas.

Some flawless stones are faceted for setting in jewellery, more often though, specimens have a magnificent array of flaws which, especially when tumble polished, produce a spectacular display of sparkling gold.

Moon stone

APOPHYLLITE

Those born with the moon in Gemini can be fickle, flirtatious, light-hearted and are often unwilling to be faithful. Apophyllite is their stone, helping to promote a persuasive tongue and good memory, while balancing opportunistic tendencies and a mistrust of their own feelings. The reconciling of emotions and intellectual ideals is the main challenge for Geminis in the game of life.

Occasionally, apophyllite has a pale pink or green hue, but the usual white or colourless form is best. These translucent crystals usually grow as closely packed clusters in cavities in basalt or limestone, and are cubic in form. Often they are scattered as if lovingly sprinkled over the surface of another silicate material.

Apophyllite demonstrates to capricious Gemini that emotion can be a beautiful thing

A soft and brittle stone, apophyllite is not particularly suitable to be worn as jewellery. The best way to display it is in a cabinet, making it a perfect talisman for Gemini because its immutable character may be seen, yet at the same time is kept at a distance – and Geminis are renowned for keeping themselves distanced from others, as well as from their own emotions. One property of apophyllite is its power to help Geminis feel more

comfortable about developing their own self-understanding. This crystal can demonstrate that for the capricious, often inconstant Gemini, expressing their emotions can be a wonderful thing .

Cancer

Key word

PROTECTION

Characteristics

TENACIOUS,
CAUTIOUS, MOODY,
NOSTALGIC

Ruling planet

MOON

Ruling stone

MOONSTONE

Sun stone

ARAGONITE

Moon metal

SILVER

Moonstone

CANCER (the crab) is ruled by the moon, which represents the three phases of female life: maiden, mother and crone. Diana, the virgin huntress, embodies the maiden and the crescent or new moon; Juno, the protector of married and pregnant women, represents the mother and the full moon; Hecate, feared and respected for her wisdom, symbolizes the crone and the dark moon. The moon is also associated with complexity and mystery.

Cancer is the sign of security. The crab is a creature well-adapted to the perilous environment of the seashore, with its constant tides, shifting sands and predators. Despite its hard outer shell and ferocious claws, the crab hides a soft interior which is easily hurt, symbolizing Cancer's vulnerable and enigmatic nature. Those born under this sign can be sympathetic and tender but they can also be emotionally insecure and timid.

Ruling stone

MOONSTONE

White stones are emblematic of the moon and moonstone is its number one gem. A semi-precious stone that seems to shimmer like liquid moonlight, moonstone can be used in any situation involving the domestic environment, mothers in general, habits, unpleasant personal memories or where Cancerian qualities are needed. Moonstone can be used to promote Cancerians' positive traits and to minimize their less desirable ones.

White moonstone (also called adularia) is a transparent, milky-hued crystal which has an

almost pearly sheen. The best moonstones have a sheen or *schiller*, caused by internal reflections, which glows softly when caught by the light.

As moonstone is turned, its irridescence moves from being invisible to bright and back again, which is a clear parallel with the lunar phases. Indeed, Pliny the elder recorded the belief that the strength of the *schiller* depended upon the phase of the moon, so that during the dark phase it could scarcely be seen.

Moonstone has a reputation for rendering its wearer invisible (particularly during the dark moon). Another tradition holds that during the waxing moon this stone produces a powerful attraction between lovers, while during the waning phase it bestows prophetic powers. It is said that if you place a moonstone in your mouth at full moon, you will be able to foretell the future – a gift of the crone.

Moonstone also has an ancient reputation for protecting travellers. Because Cancer is associated with the stomach, moonstone can be helpful for dieters and may also ease the suffering of people with eating disorders such as anorexia or bulimia.

Sun stone
ARAGONITE

Cancer is a complicated sign and those born under it often experience conflict. On the

*B*ecause Cancer is associated with the stomach, moonstone can be helpful for dieters and people with eating disorders such as anorexia or bulimia

*A*ragonite promotes caution and balances the Cancerian tendency to be over-exacting with others – especially children

one hand they have an innate need for self-preservation; on the other, they are often overwhelmed by sheer curiosity. This means that Cancerians may often be shy and introverted yet are constantly buffeted by the urge to explore and take full part in life. Unfortunately, they are also prone to the legendary Cancer moodiness!

A strong tendency to allow emotion to cloud their thinking – and, more dangerously, their decision-making – can lead to a very subjective, self-centred view of life. At their worst, Cancerians are self-pitying, erratic in mood and action, easily flattered and suffer from an inferiority complex.

Crabs do not like to 'let go' of anything, from feelings and habits to people. At their best, Cancerians are nurturing and cherishing. However, those who suffer from a chronic lack of self-confidence may turn to emotional blackmail, which eventually leads loved ones or partners to resent and despise them.

The fundamental stone for these Cancerians is yellow aragonite. This promotes caution and balances the Cancerian tendency to be over-exacting with others – especially children. It also reduces the urge to resort to emotional blackmail.

Like Cancerians themselves, aragonite is not the best of stones to handle: therefore it is best kept on the desk or by the bed – where it can be seen and occasionally touched.

The colourful nacre or mother-of-pearl on the inside of seashells is aragonite, as are the

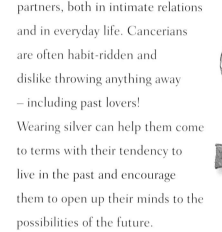

Silver jewellery may be worn as an overall protection against the trials of life

skeletons of many other marine organisms, intimately linking the stone with Cancer.

Moon metal
SILVER

People born with the moon in Cancer tend to be sentimental. They are also susceptible to sensual stimulation – muted lights, soft music, silk, fur or velvet against the skin, all of these can work magic for Cancerians. They may however, become overly dependent on partners, both in intimate relations and in everyday life. Cancerians are often habit-ridden and dislike throwing anything away – including past lovers! Wearing silver can help them come to terms with their tendency to live in the past and encourage them to open up their minds to the possibilities of the future.

Silver has been associated with the moon, and hence with Cancer, for thousands of years. This white metal (second only to gold for malleability) rapidly tarnishes, its surface turning black. The transformation from shining white to pitch black reveals a remarkable parallel with the moon's phases.

Dreaming of this metal is said to be a portent of financial satisfaction and silver is still popular as a talisman for attracting good luck and warding off misfortune. It represents the ability of humanity to foresee and therefore elude danger and silver jewellery may be worn as an overall protection against the trials of life.

Leo

Key word
CREATIVITY

Characteristics
**NOBLE, BIG-HEARTED,
GENEROUS, AUTOCRATIC**

Ruling planet
SUN

Ruling stone
TIGER'S EYE

Sun metal
GOLD

Moon stone
DIAMOND

Tiger's eye

LEO (the lion) is ruled by the sun, personified by Apollo, god of light, music, civilization and the self and a positive and stimulating character. The sun embodies willpower, the drive to achieve great things, dynamic creativity and nobility of character.

Leo is the sign of creation and the lion is preoccupied with the urge to create something – anything – that is uniquely its own. Leos often ignore detail in favour of the larger view, forgetting that others are not extensions of their own egos. The ultimate challenge for Leos is to develop their own self-esteem from vast, but often unrealized, inner resources. A Leo can become as much the centre of its personal world as the sun, its ruler, is the centre of our solar system. The danger in this is that Leo can bask lazily in others' admiration without generating its own nurturing, enlightening power from within.

Ruling stone
TIGER'S EYE

Yellow stones are emblematic of the sun, and tiger's eye is its primary stone. Twinkling and semi-precious, it is a fabulous embodiment of nature's flamboyance. Tiger's eye can be used in any situation involving the self, wellbeing, fitness, and fathers in general, or wherever Leonine qualities are needed. At their best, those born under the sun sign of Leo have creative vision, expansiveness, unconquerable willpower, benevolence and magnanimity. They can also be domineering, hedonistic, indolent, arrogant and lacking in self-confidence. Tiger's eye can be used to promote their positive traits and to minimize their character flaws.

Roman soldiers wore engraved tiger's eye for protection in battle. There were two reasons for this custom: tiger's eye was thought to steady the nerves, and because it was said to beguile the eye, the stone – and by association the soldier – became a hard target to hit.

> *Tiger's eye can be used to promote attention to detail and a willingness to listen to others' points of view*

Tiger's eye promotes a deeper understanding of our own inner resources. As a meditation stone, it is especially useful when contemplating dualism. Turn a polished example of tiger's eye in the light and its dull brown bands are transmuted by the alchemy of chatoyancy into glowing gold, clear and bright. Tiger's eye stones are often found in rings, which are ideal for Leos. Tiger's eye can be used to promote attention to detail and a willingness to listen to others' points of view.

Sun metal
GOLD

Gold has been associated with the sun and with Leo since time immemorial. This bright yellow metal occurs in a number of crystal forms, notably cubic, or as nuggets or, most commonly, as tiny flecks. Gold is the heaviest natural mineral on earth and is so ductile that a small pile of gold dust can be turned into a solid disc with a single hammer blow. Pure gold is also so soft that even a thumbnail can dent or scratch it but it is a remarkably stable element that will neither corrode nor tarnish.

Leos often lack the self-assurance, boundless self-confidence and self-possession for which they are famous. Lions need to be needed, otherwise what are they to do with all

> *Because gold is so incorruptible, it is used as an overall healing stone to promote good health and wellbeing*

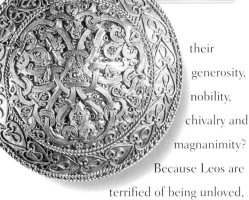

their generosity, nobility, chivalry and magnanimity? Because Leos are terrified of being unloved, unappreciated and unwanted, wearing gold can act as a symbol of the status they feel they deserve and to reassure them of their own worth.

Although it has fascinated goldsmiths for more than 4,000 years, this mineral's very softness rendered it unsuitable for anything but ornamental use for much of this time. It was first used as currency in the 6th century BC by Croesus, king of Lydia, Asia Minor and that remained its primary role until recently when its inherent properties earned it a place in electronics.

Primarily due to its colour, gold is reckoned to be a masculine metal. Its resistance to corruption and the sun's daily (and annual) return to vigour link the metal strongly with notions of immortality, resurrection and reincarnation.

The druids cut their sacred mistletoe (also known as 'All Heal') with a sickle of gold – a tradition that revealed their intention to harvest the herb without 'killing' its magic.

Because gold is so incorruptible, it is most often used as an overall healing stone to promote good health and wellbeing.

The strength of diamond makes it useful for overcoming danger

feelings and emotions as well as with those of others. Diamond is the stone for these Leos and it can also help to balance those less worthy Leo traits of conceit and arrogance.

The hardest mineral, diamond derives its name from a Greek word meaning invincible. A real diamond will scratch glass, although so too will many minerals that the unscrupulous might seek to misrepresent.

The hardness of diamond is interpreted as symbolic of enduring love, but in early times its toughness made it talismanic of manliness and it was taboo for a woman to wear it. According to tradition, putting a diamond in your lover or spouse's hair when they are asleep on their back can determine their faithfulness. He or she will turn toward you if they are faithful, away from you if not!

The strength of diamond makes it useful for overcoming danger and its durability means it can help cut through delirium, muddled thinking and cure insomnia, making it perfect for big-hearted but sometimes bewildered Leos.

A dream about diamonds can mean two different things: if you own one, it predicts a minor financial loss, if not, expect a modest financial gain.

Moon stone

DIAMOND

Those born with the moon in Leo are usually extravagant (to the point of being overwhelming), dignified and confident. Their affections are wholehearted and generous and lions will go to extraordinary lengths for friends and loved ones. Leo is one of the luckier signs and lions are generally uninhibited and comfortable with their own

Virgo

Key word
CLASSIFICATION

Characteristics
**PRACTICAL, MODEST,
PRECISE,
SELF-CONSCIOUS**

Ruling planet
MERCURY

Ruling stone
ROCK CRYSTAL EGG

Sun stone
**PETRIFIED OR
OPALIZED WOOD**

Moon stone
**STILBITE,
MILKY QUARTZ**

Rock crystal egg

VIRGO (the virgin) is ruled by Mercury. Mercury was the messenger and god of communication and his intellectual and rational character enhances the logical and realistic nature of Virgo. However, it has been suggested that the planet Chiron (discovered in 1977 orbiting between Saturn and Uranus) might come to be considered Virgo's ruling planet. Chiron was a healer and Virgo's secondary urge is to be of service to the world, disseminating the data it has gathered to others.

Virgo is the sign of classification. Virgos are always acquiring information and have a compulsion to understand what is happening in their own lives and in the world at large. The symbol for Virgo, the corn maiden, represents the harvest, not only of grain but of ideas. Virgo people are discriminating, analytical and adept at assimilating information. At their best, those born under Virgo are conscientious and have a logical and systematic approach to life. On the down side, they can be stubborn, pedantic and can possess a great fear of the unknown.

Ruling stone
ROCK CRYSTAL EGG

Rock crystal – in all its forms – is symbolic of Mercury and the rock crystal egg is this sign's important stone. The egg is a powerful symbol of latent potential, undiscovered destiny, and a safe haven during a time of extreme vulnerability. It contains the promise of new life. It can be used in any situation involving precision and attention to detail,

service of any kind (whether received or rendered) and wives in general, or where Virgoan qualities are needed. It helps to promote the positive traits and to counteract the flaws in the Virgo character.

The smooth, comforting shape of the 'worry egg' helps ease stress and calms an overactive mind. The egg is the emblem of Eostre, the Teutonic goddess of spring, from whose pagan festival we derive the name for Easter. If the rock crystal egg is clear of impurities, so much the better, but 'rainbows' (caused by the refraction of light from internal flaws), are actually welcome and are a delight to the eye.

> *The smooth, comforting shape of the 'worry egg' helps ease to stress and calm an overactive mind*

The egg is useful as a meditation stone and you can turn it over in your hands as you turn over a problem in your mind. The bulbous end represents the big picture, the full array of considerations, causes and effects that are involved. The point of the egg represents the personal impact of all this on you (or whoever/whatever is the subject of your conjecture). This ability to simultaneously connect with both the general and the particular can help us to grasp the precise implications of even subtle changes to our environment. Such understanding of how the future may shape our lives gives us the wisdom to take charge of our own destiny, rather than be at the mercy of unknown forces.

Sun stone
PETRIFIED
OR
OPALIZED
WOOD

Virgos are often caricatured either as pernickety perfectionists with a horror of dirt, or as eternal secretaries. Although there is a grain of truth in these depictions, it is a very small grain indeed. Virgos are essentially refined characters who have a horror of disorder. They don't worry about dirt: Virgo is an earth sign and those born under it are pragmatic and accept the natural order of things.

Virgo people are quite capable of forgetting all about the housework while they delve deeply into their own minds. There may be an interest in hygiene but it will be related to the growth and effect of bacteria rather than what the kitchen looks like! The only perfection a Virgo insists upon is his or her own understanding of the facts. The Virgoan need for efficiency is often mistaken as perfectionism but it is really a pragmatic attitude: efficiency is less time-consuming and more effective than sloppiness, either at work or mentally, and is very much a Virgo ideal.

The fundamental stones for those born under the sun sign of Virgo are petrified or

> *Petrified wood can be used as a charm to increase longevity*

opalized wood. These bolster Virgoan tendencies for impartiality, modesty, lucidity and attention to detail, while balancing their lack of enthusiasm and puritanical and hypercritical tendencies.

Petrified wood, also known as fossilized wood and opalized wood, are two forms of the same thing, although the opalized form is rarer. To become a fossil the tree must first avoid the natural cycle of disintegration through decay, which usually occurs when it is rapidly submersed in water, silt or mud and so eventually forms sedimentary rock. Later, mineral-rich solutions permeate the sediment and silica replaces the original wood, forming a replica so detailed that even the plant's cell-structure can be preserved.

The sheer age and intricate detail of these stones will appeal to Virgo's logical mind, and, talismanically, allow them to explore the reaches of their own psyches. It may also act as a reminder that too rigid a mental structure can result in a static, stagnant mind, much as a once-living, breathing, growing tree has been changed into stone.

Traditionally, petrified wood has been used to ward off negative forces. These days it can be used as a charm to increase longevity. Another popular use for this relic of ancient life is as a touchstone. This has many uses – it can help you to cast your mind back before birth, to investigate previous incarnations, or to explore some of the other diverse life forms on our planet through the aeons – wonderful subjects for Virgoan investigation!

Moon stone

STILBITE, MILKY QUARTZ

Those born with the moon in Virgo are fastidious, yet incline towards emotional dishonesty, a fear of losing control and a tendency to find fault. Virgos may be either refined, conventional and reserved, or so afraid of losing control that everything (almost literally) comes from a manual. Emotional and intimate relationships cannot be neatly pigeonholed – not in the real world, anyway – and are therefore to be feared. Virgos can sometimes take this to extremes and end up existing in a cold, self-imposed loneliness, which is a terrible pity, because the world needs a little more of the gentility and finesse that Virgo can bring to partnerships.

The primary stones for these Virgos are milky quartz and stilbite (a member of the zeolite group). Milky quartz has long been worn by mothers to ensure or promote lactation. In chronic illnesses where a patient cannot take solid food, this stone has a reputation for promoting swift recovery and steady recuperation. Stilbite resembles a freshly harvested wheatsheaf and as such reminds Virgos of their earthy heritage. The stone encourages Virgos to develop and nurture their many talents in order to help both themselves and others.

Stilbite encourages Virgos to develop and nurture their many talents in order to help both themselves and others

Libra

Key word

BALANCE

Characteristics

DIPLOMATIC,
TACTFUL, PEACE-LOVING,
INSINCERE

Ruling planet

VENUS

Ruling stone

CHRYSOPRASE

Sun stone

PYRITES

Moon stone

SELENITE

Chrysoprase

L IBRA (the scales), is ruled by Venus (although it has been recently suggested that the sign's true planet remains unidentified). Peace-loving Venus was a fertility goddess, an aspect of the Great Mother, who ruled over the natural world. Her Norse counterpart was Frigg, the wife of Odin. Frigg's foresight and wisdom matched her husband's and often prevailed when their intentions differed – although, like Libra, her intelligence and astuteness did not conflict with her love of jewellery and beautiful things. Venus fosters an instinctive understanding that life cannot exist without the interaction of relationships, however painful they may sometimes be.

Libra is the sign of harmony and those born under this sign strive for fairness, equity and balance in all things. At their best, Librans are diplomatic, equable and just. On the other hand, they can also suffer from insincerity and vacillation. Chrysoprase can be used both to promote the positive traits, and to minimize or balance the less desirable qualities.

Ruling stone

CHRYSOPRASE

Green stones are emblematic of Venus and chrysoprase is Libra's ideal stone. This lovely, spring-green coloured gem is emblematic of the apple, symbolic of health and the knowledge of good and evil. Chrysoprase can be used in any situation where diplomacy and tact are required, such as in meetings and dealings with acquaintances or workmates, or where Libran qualities are needed. Apart from its soothing colour and texture, chrysoprase is frequently patterned with opaque creamy veins and, occasionally, minute quartz crystals glitter in cavities like a myriad tiny stars. These charming features remind us of the possibilities inherent in the chance encounters that destiny may spring upon us.

Chrysoprase is a 'feel-good' stone – touching it can promote feelings of happiness and of being at one with the world

Chrysoprase is an apple-green variety of chalcedony, a translucent, white mineral. Chalcedony has many natural coloured varieties and, because it is porous, has been subjected to a wide range of chemical dyes to produce colours to order. Many of these artificial varieties soon lose their colour, and are best avoided.

In the Middle Ages chrysoprase was reckoned to render anyone who carried it in their mouth invisible. A Romanian tradition held that its owner could comprehend the language of lizards. Engraved with the figure of a bull, chrysoprase was employed as a powerful protective amulet during the 13th century. It was also reputed to lose its colour if poison was present.

As a talisman for Librans, chrysoprase acts to focus the mind on the physical aspects of life and promotes enjoyment of simple pleasures: a walk in the sunshine; the fragrance of flowers and herbs; the taste of sweet fruit; a warm, comforting hug…

Chrysoprase is a 'feel-good' stone – touching it can promote feelings of happiness and of being at one with the world.

Sun stone

PYRITES

Those born under the sun sign of Libra strive to bring goodness, truth and beauty into the world – although being Libran does not automatically mean that the individual is able to be and do these things! More often than not it is a very great struggle indeed to embody and express any of these ideals and this is Libra's great life challenge. Fortunately, most Librans have the objectivity and wisdom to analyze and blend opposites into a harmonious whole, which makes their task easier.

The sun stone for Librans is pyrites, also called iron pyrites or fool's gold. It helps promote eloquence and persuasiveness and balances a tendency to be easily distracted. Pale brass or golden-coloured, pyrites have frequently been mistaken for gold by over-optimistic prospectors, and to add to the confusion, the two minerals often occur in the same place. But pyrites does not share gold's incorruptible qualities – it corrodes in the presence of air.

The heavy solidity of pyrites helps bring Librans back down to earth or 'ground' them

Fragile specimens, such as some pyritized fossils, created where pyrites replace the shells of marine organisms for example, can decompose within a matter of months. Often occurring as striated, intersecting cubes, pyrites are always opaque and are widespread, occurring in igneous, sedimentary, and metamorphic rock. Pyrites have been used to create sparks, and hence start fires, since prehistoric times. The name pyrites is derived from the Greek word for fire and the stones were sometimes employed instead of flint in early firearms.

Pyrites can be polished to produce a mirrored surface, a technique used in ancient Mexico. The ancient Chinese believed that it had protective properties and used it to ward off alligators.

The best talisman for Librans is a single pyrite cube, as large and as perfect as possible. Librans tend to be, mentally, a million miles away from earth and very preoccupied with cerebral matters. The heavy solidity of pyrites helps bring Librans back down to earth or 'ground' them, while its mathematically attractive shape and appearance appeals to them on a higher level, too.

Moon stone
SELENITE

Those born with the moon in Libra may find their personal life is often difficult because of an innate desire for stability. Of course, when the moon is full it is on the opposite side of the earth to the sun, a powerful image for Librans, who delight in their sense of cosmic balance.

While essentially peace-loving, Librans are also very aware of the faults of others and their inability to do anything to alleviate these faults can lead to emotional stress. In the main, however, these Librans are friendly and charming, although perhaps sometimes a little too eager to please others. They are often gentle, refined and discriminating, but on the negative side they can also be lazy and rather evasive.

Their fundamental stone is selenite, a transparent, colourless or white crystalline variety of gypsum. Crystals occasionally twin, producing a fishtail, or swallowtail formation. This type of stone was named in honour of the Greek moon goddess, Selene, who particularly personified the full moon and whose brother was Helios, the sun.

Selenite has a powerful reputation for smoothing the path to reconciliation for estranged lovers. Ideally, you should try to find a large piece of selenite and keep it beside the bed, preferably where the light can shine on and through it. Talismanically, selenite may help Librans to be more honest about their own emotions.

Talismanically, selenite may help Librans to be more honest about their own emotions

Scorpio

Key word
CONTROL

Characteristics
**INTENSE,
COURAGEOUS,
INCISIVE,
POSSESSIVE**

Ruling planet
PLUTO

Ruling stone
SERPENTINE

Sun stone
SULPHUR, GOLDEN BERYL

Moon stone
WHITE OPAL

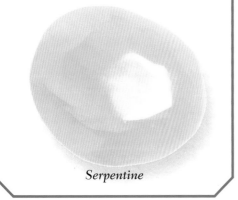

Serpentine

SCORPIO (the scorpion), is ruled by Pluto, which represents the dominion of the Underworld, and extends to the darker reaches of the human psyche. Pluto was lord of the Underworld and all its treasures, which can be seen as the powerful personal resources to be found in the human mind and the subconscious. Pluto's sombreness and understanding of the dark side of life provides a good insight into Scorpio's powerful, labyrinthine character.

Scorpio is the sign of power and is probably the most complex and least understood of the zodiac signs. There are three distinct Scorpio types. The first and least developed type of Scorpio is symbolized by the scorpion. Scorpions use their talents for purely selfish ends, bear grudges and work in secret to repay insults many-fold, regardless of the fact that they may be emotionally crippled or killed in the process!

The second type is symbolized by the mysterious dragon. Dragons hoard their treasures and secrets and live a solitary life, often tempting others to see them as challenge and try to steal their prizes.

The third and most highly developed type of Scorpio is symbolized by the eagle. The high-flying eagle possesses keen eyes that see everything and uses its talents with compassion to improve both its own life and the life of others.

At their best, Scorpio people are courageous, passionate and intense. At their worst they can be ruthless, suspicious and extremely possessive.

Ruling stone
SERPENTINE

Patterned stones are emblematic of Pluto, and serpentine is Scorpio's primary stone. Intricate, beautiful, brittle yet versatile, this stone symbolizes both the complexity of Scorpio and the uncompromising severity of Pluto. It can be used in any situation involving conservation, change, stress, bereavement, nightmares, mental health or where Scorpio qualities are needed. Serpentine can also be used both to promote Scorpios' positive traits and to minimize their less desirable ones.

Serpentine 'worry eggs' are particularly helpful as their hard, cool smoothness and comforting weight can alleviate stress

Serpentine is so called because its streaked and spotted patterns are fancifully held to resemble snakeskin. This traditional association with the serpent, an emblem of death, power, and occult knowledge – its ability to slough its skin is a widespread metaphor for reincarnation – is reinforced by contemporary crystal mystics. It is popularly used to promote an intuitive understanding of the natural cycle of birth, death and rebirth, so bringing about inner peace, strength and security and an awareness of the need for conservation.

Serpentine is almost always opaque and has a wide variety of patterns, ranging from criss-crosses to swirls. Colours, too, can vary considerably, from pale green to deep red, perhaps veined or blotched with white or black. The best stones for Scorpios are those that exhibit clear contrasts in their patterns. Serpentine 'worry eggs' are particularly helpful as their hard, cool smoothness and comforting weight can help alleviate stress.

Extensively used for ornaments (such as candleholders, pots, bowls and bookends), highly prized green and red serpentine can be found at the appropriately named Lizard Point in Cornwall, England.

Sun stones
SULPHUR, GOLDEN BERYL

Those born under the sun sign of Scorpio are incisive and strong enough to stand firm for personal beliefs, while having a tendency toward possessiveness and ruthlessness.

Make no mistake, Scorpios are not to be taken lightly. They have a realistic appreciation of the world in which they live but veer toward the darker side of life. They face their own shortcomings, dark fantasies, lusts and nastiness with the same fidelity that they bring to life as a whole. No other sign has such profound personal integrity.

Scorpios find it difficult to express compassion because very often they have had to struggle against their own frailties throughout life and so have little time for weakness in others. Nor will their pride, self-control and self-containment allow them to ask for, let alone accept, help from their fellows. In their personal lives Scorpios find it almost impossible to forgive themselves for their own failings. They succeed through their own efforts and feel that everyone else should do so too!

The fundamental stones for Scorpios are sulphur and golden beryl. Sulphur, also called brimstone, is made up of translucent, bright yellow crystals that are very delicate and must be treated with care – even the warmth of your hands can make them expand enough to crack. This means that the stone needs to be kept as dry as possible and handled only rarely.

Sulphur is found wherever there is volcanic activity. It was sacred to Athene, the Greek goddess of wisdom, and so symbolizes Scorpio, wise in the ways of the world, very well.

The beryl family includes emerald, aquamarine and morganite, but only golden beryl is appropriate here.

Beryl has been used as a divinatory and healing stone for many years. Its ancient name was heliodor, meaning 'gift of the sun', and worn as jewellery it acts talismanically to bring light into Scorpio's shadowed soul.

Moon stone
WHITE OPAL

Those born with the moon in Scorpio can be selfish, hold grudges and may also suffer from periods of depression. The influence of the moon in Scorpio helps balance sadistic

The sunny sulphur crystals help to illuminate the dark corners of Scorpio's psyche.

White opal is considered beneficial, especially for a weak heart, and is held to ward off airborne diseases

tendencies and extreme possessiveness.

Scorpios are very sensual, often have magnetic sex appeal and need to be in control of their own feelings, their lives and circumstances – which is where the image of the power-mad Scorpio comes from. Unfortunately, the moon's influence on the sign can be very negative, having the effect of chipping away at Scorpio's self-control and self-confidence.

The stone for these Scorpios is white opal, also called milky opal or common opal. Close attention will reveal the subtle beauty of white opal, which is a solidified jelly containing as much as 10 per cent water in minute spaces between the tiny globes of silica that form the mineral's structure. The skeletons of some marine organisms, including sponges, are composed of opal.

Because opal is fragile, fractures form easily and can cause irreparable damage. Grease can also get into the stone and will dull its colour, so handle with care.

Opal is an ideal talisman to counteract the negative effects of the moon in Scorpio. In Roman times, it was known as Cupid's stone because of its association with love. In the Orient, opal was believed to be the anchor of hope. It is considered beneficial for a weak heart, and is held to ward off airborne diseases. It has retained its Roman reputation for enhancing the attractiveness of its wearer by bringing out their inner beauty – making it perfect for Scorpios.

Sagittarius

Key word

JUDGEMENT

Characteristics

**COMPASSIONATE,
OPEN-MINDED,
ENQUIRING, SEARCHING**

Ruling planet

JUPITER

Ruling stone

SAPPHIRE

Sun stone

RUTILATED QUARTZ

Moon stone

WHITE CALCITE

Sapphire

SAGITTARIUS (the archer) is ruled by Jupiter, ruler of heaven's host. The morally enlightened Jupiter was chief god of the Roman pantheon, the guardian of the law and the giver of justice. He symbolizes conscience, consciousness and the urge to expand human understanding.

Sagittarius is the sign of wisdom and is symbolized by a centaur aiming an arrow to the stars. This image represents humankind struggling to overcome its animal nature and achieve civilization and, eventually, enlightenment. At their best, Sagittarians are kind, academic, optimistic, altruistic and aspirational. However, they may also be over-confident, extremist, boastful and dogmatic.

Ruling stone
SAPPHIRE

Blue stones are emblematic of Jupiter and sapphire is Sagittarius' foremost stone. This beautiful, sky-blue, precious gem symbolizes the 'vast beyond' that Sagittarians seek to explore. Sapphire can be helpful in any situation involving legal and financial matters, education and academic concerns, teachers and instructors in general, or where Sagittarian qualities are needed. Sapphire can be used to promote Sagittarians' positive traits and to balance their less desirable qualities.

Sapphire is a blue variety of corundum. The finest gem quality stones are said to be cornflower blue but any stone that appeals to you is suitable. It is not even necessary to possess one of the transparent crystals; sapphire occurs in a granulated form that is

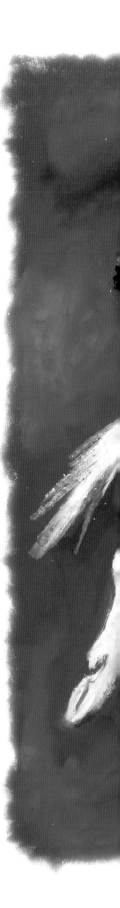

quite unsuitable for use in traditional jewellery and is therefore relatively inexpensive.

In the 12th century, Pope Innocent III decreed that bishops' rings should be set with sapphire because it signified chastity. Being unusually cool to the touch, sapphires were reputed to quench fire and, by analogy, the fiery human passions such as lust and anger. They also represented the yearning to escape from the mundanities of life into a celestial paradise. Buddhists, too, held that the jewel assisted the spirit to be moved to devotion and a pure life.

For those seeking to right the world's wrongs, sapphire has a reputation for being able to free the individual from captivity and provide protection against their enemies' plots.

Again in the 12th century, it was attested that you could kill an adder – then regarded as an agent of diabolical forces and an enemy of truth and justice – by placing a sapphire in its mouth. Deep-coloured specimens are still prized as being able to cleanse the mind of poisonous thoughts and baseless assumptions.

Sapphire also traditionally confers the gift of prophesy and is thought to promote a clear and true vision of the future, unwarped by

Rubbing the eyes with sapphire is reputed to ward off physical as well as spiritual blindness

prejudice or guile. Sapphire is currently held to ease the projection of the astral or etheric body (and even assist psychokinesis), by preventing illusory and dangerous distractions. Rubbing the eyes with this stone is also reputed to ward off physical as well as spiritual blindness (as an eyewash the stone needs to be immersed in the water both before and after the eye is bathed).

Sun stone
RUTILATED QUARTZ

Those born under the sun sign of Sagittarius are honest, charitable and young at heart. On the negative side, they can be careless and overconfident. There are two sides to Sagittarius : one is sporty, outdoor-loving and a lover of the simple things in life; the other is an eager, curious student, thirsty for knowledge and a delight to teach. Both are often found together.

Sagittarians are usually restless, always seeking to expand their horizons. They are the sort of people who inspire you to give up everything and trek overland to Tibet in order to sit at the feet of a guru. Yet at the same time, Sagittarians can be very conservative. Although they are always open to new ideas and experiences, they also need to know that there is somewhere safe and familiar to return to, even if, by the time the archer has imported all its new knowledge, the place is no longer quite so familiar!

The stone for those born under the sun sign of Sagittarius is rutilated quartz, also called Cupid's arrows, maiden's hair or sagenite. Here, clear quartz contains inclusions of needle-like crystals of rutile (titanium oxide). Depending on their quantity and arrangement in the quartz, these long, minutely thin golden crystals can resemble blonde hairs, shafts of sunlight or glistening arrows.

There is something almost romantic about the attraction of this stone. Sometimes a tumble-polished specimen of palest smoky quartz can have rutiles so fine that they can only be seen by turning the stone in the light – when they seem to glisten like angel's hair. Because romance and remaining young at heart are held to prevent the onslaught of premature ageing – or perhaps simply because these 'blonde hairs' never turn grey – rutilated quartz is thought to boost energy, youthfulness and health and promote all those bodily defences that stave off the assault of advancing years, something that seems to come very naturally to typical Sagittarians.

Rutilated quartz is thought to boost energy, youthfulness and health

The clarity of white calcite may help Sagittarians find the most useful ways of sharing their knowledge and wisdom with others

Moon stone
WHITE CALCITE

Those born with the moon in Sagittarius are independent, sympathetic and have inquiring minds. They may also be overly subjective, argumentative, and greedy. In the main, Sagittarians are open and honest souls with a tendency to support large-scale charitable efforts – and also to make sweeping statements about subjects under discussion!

Their stone is white calcite, also called calc spar, nailhead spar, dogtooth spar or beef. (For optical calcite, see Aquarius moon stone). Specimens range from transparent crystals to translucent white granular forms, all or any of which can be used here.

Calcite has more different crystal shapes than any other mineral. Building up a collection of their own and discovering the various properties of calcite will particularly appeal to the Sagittarian thirst for knowledge. Colourless when pure, specimens of every hue can be found, including black. Calcite forms the skeletons of most marine organisms, some of whose remains accumulated in vast drifts and were fossilized, forming limestone, chalk and marble. Stalagmites and stalactites are formed from dissolved calcite deposited when drops of water evaporate.

The clarity of white calcite may help Sagittarians find the most useful ways of sharing their knowledge and wisdom with others, by acting as a focus for logical thought and decisive action.

Capricorn

Key word
STATUS

Characteristics
DEPENDABLE,
HARDWORKING, AMBITIOUS

Ruling planet
SATURN

Ruling stone
JET

Sun stone
YELLOW JASPER

Moon stone
WHITE CHALCEDONY

Jet

CAPRICORN (the goat), is governed by Saturn. The exacting and onerous Saturn was originally an agricultural deity who came to be seen as the founder of civilization and social order. He was also associated with Kronos and the Egyptian Seb (god of the earth) whose name means 'time'. Eventually, Saturn came to be envisaged as a gaunt figure with a scythe and hour-glass, symbolizing Time itself, which ends in death.

Capricorn is the sign of integrity. It represents all the trials and hardships of life which serve to refine the mind and the spirit into a powerful, potent unity. It is also the sign of the patriarch, which means that Capricorns can sometimes be heavy-handed. At their best, Capricorns are prudent, ambitious, economic, self-disciplined and diligent. Their unattractive qualities are meanness, snobbery and severity; they also have morbid tendencies and are predisposed to using others to further their own ends.

Ruling stone
JET

Black stones in general are emblematic of Saturn, and jet is Capricorn's stone. An intensely black stone, jet symbolizes the harshness of Saturn. It can be used in any situation involving labour, tests, examinations, doctors, managers or enemies or when Capricorn characteristics are needed. Jet can be used both to promote Capricorns' positive traits and to minimize the less desirable ones.

Jet is an organic stone – a dense variety of lignite. It is formed from fossilized wood and

is newer than coal, but more ancient than peat. Jet is fairly tough and yet is easily carved. It has been extensively used in Britain since neolithic times: there is a notable outcrop in the coastal cliffs at Whitby, North Yorkshire.

Jet was worn in ancient times to honour the Phrygian goddess of the earth, Cybele, whose son Midas had the golden touch and laced the river Pactolus with gold dust – a nice analogy of the treasures that Capricorn's hard work can achieve. Along with other deities, Cybele was subsumed into the Greek goddess Demeter, who presided over the fertility of the earth, and some gardeners and farmers still wear jet in order to give their crops the best chance in life.

It has been used in the same ways as amber (see Taurus sun stone) and it shares another characteristic: when rubbed it gains an electrical charge. Also like amber, it can be used as an incense. Sometimes this smoke is used for scrying and in the past it has also been used to banish demons.

Jet has been used in the manufacture of fancy goods such as brooches, bracelets and beads – including, during the 16th century, rosary beads – and practical items such as buttons, knife handles and belt fasteners were also popular. Occasionally statuettes were fashioned, including chess pieces. Its association with 'widow's weeds' came about because Queen Victoria wore the stone as

Jet is best worn as jewellery to promote strength of character and determination in the face of adversity

part of her perpetual mourning costume. Jet is best worn as jewellery to promote strength of character and determination in the face of adversity. As Capricorn is associated with the skin, wearing it next to the skin is most beneficial. Jet beads and rings, especially those that have no backing to the setting so that the stone touches the flesh of your finger, are particularly appropriate.

Sun stone
YELLOW JASPER

Those born under the sun sign of Capricorn are generally dependable and straightforward, although there is a tendency toward meanness and social climbing.

Capricorns often seem to be much older than they really are when young, because their early lives are frequently plagued by family problems and they often have to 'grow up' before their time. Capricorns also have an overall need to conform to society's values, rules and regulations, which accords with the sign's patriarchal nature. Capricorn is the sign that embodies the 'father' in the old-fashioned sense – as the head of the household whose word is law.

The stone for sun-born Capricorns is yellow jasper, also called leontios. This unusual variety of chalcedony often has fascinating swirls of light and dark shades on its surface that can be seen at their best in tumble-polished stones. Jasper was one of the stones used in the foundations of

Carrying tumble-polished jasper will help to ease the strain for Capricorns when others are relying on them

the New Jerusalem. Engraved with the figure of a lion or an archer, jasper was once used to lower fever, calm the overwrought imagination and lessen the risk of poisoning. It has also accrued a strong reputation for being able to protect the wearer from violence and is renowned for staunching the flow of blood from wounds or during menstruation. It also makes pain easier to bear, especially that of childbirth. Indeed, jasper is thought to facilitate all births, whether human, animal or conceptual.

Jasper's association with foundations makes it the ideal talismanic stone for Capricorns who, because they strive for conformity and competence in all they do, are often seen as utterly dependable. Such a reputation can be difficult to live up to. Carrying tumble-polished jasper will help to ease the strain for Capricorns when others are relying on them.

Moon stone
WHITE CHALCEDONY

Those born with the moon in Capricorn are diligent, self-disciplined and have a serious attitude to life and personal relationships. They may also be

self-doubting, materialistic and mercenary. Their stone is white chalcedony, a translucent mineral that has many natural coloured varieties. Because it is porous, white chalcedony has also been subjected to a wide range of chemical dyes to produce colours to order. Many of these artificial varieties lose their colour, especially when exposed to sunlight, and are generally to be avoided.

Chalcedony is named after the ancient Greek city of Chalcedon. It is a cryptocrystalline variety of quartz, with crystals so minute that they are invisible to the naked eye. Specimens are easily spoiled by staining liquids, so take care not to spill anything over them.

White chalcedony is held to promote the secretion of milk, to encourage chastity, dispel phantoms, and cure depression, the latter being a typical problem of this sign. Capricorns tend to take themselves very seriously, which at worst, may lead to an inflated sense of self-worth and bitter resentment that others do not see things the same way. Chalcedony's bright, smooth whiteness encourages Capricorns to see the lighter side. Keep a piece on the desk or carry some with you when life seems grim.

> *White chalcedony's bright, smooth whiteness encourages Capricorns to see the lighter side. Keep a piece on the desk or carry some with you when life seems grim*

Aquarius

Key word
ENLIGHTENMENT

Characteristics
**INNOVATIVE,
INDIVIDUAL, SUPRISING**

Ruling planet
URANUS

Ruling stone
KUNZITE

Sun stones
**BROWN ZIRCON,
CHIASTOLITE**

Moon stone
OPTICAL CALCITE

Kunzite

AQUARIUS (the water-bearer), is ruled by Uranus, which symbolizes the night sky. Uranus's union with Gaea, the Earth, resulted in the births of the Titans, among whom was Kronos (Saturn). The eponymous planet's discovery in 1781 coincided with the dawn of the current technological age sparked by the industrial revolution. Uranus became known as the planet of innovation and new discoveries, and is now representative of humankind reaching for the stars.

Aquarius is the sign of knowledge – but not, alas, of wisdom! Aquarians are single-minded in their devotion to following ideas through to the bitter end, regardless of the consequences and usually without taking the time to check the possible repercussions. Aquarians can be individualistic, eccentric, intellectual and given to flashes of inspiration and insight. On the other hand they can also be irresponsible, anarchic and suffer from feelings of alienation.

Ruling stone
KUNZITE

Purple stones are emblematic of Uranus and kunzite is its primary stone. This fascinating, dichroic gem symbolizes the innovation and new discoveries that are the particular domain of Aquarius. Kunzite can be used in any situation involving new discoveries or inventions, secret activities, friends, strangers, or where Aquarian qualities are needed. Kunzite can be used both to promote the positive traits of Aquarius and to balance the less desirable qualities. Kunzite, although a relative newcomer to the treasury of helpful crystals and other minerals, has already gained a prodigious reputation. In particular it is renowned for aiding relaxation, acting rather like a lightning rod that defuses potentially lethal amounts of frustration. Such properties are essential in coping with the powers unleashed by Aquarius.

Kunzite is a transparent, lilac or purple form of spodumene, (which is usually white), and was first found in California in 1902. Its occult strength lies in its dichroicism: when looked at side on, the mineral is pale in hue, yet viewed along its axis, the colour is much richer and darker. The way its appearance changes when observed from different angles perfectly illustrates how a mere alteration of viewpoint may be all that is needed to see a whole new approach to problems.

Kunzite is also said to be beneficial in regulating the menstrual cycle, thereby enhancing a woman's quality of life. It can be used to foster self-discipline and a sense of responsibility. Although kunzite can be worn as jewellery, exposure to sunlight can fade its colour. To keep it in a pristine state, carry a small piece or keep it near you at work, so that you can handle it from time to time.

Sun stone

BROWN ZIRCON, CHIASTOLITE

Those born under the sun sign of Aquarius can display detachment and individuality, while being attracted to the unusual and even the bizarre, especially where sexual practices

Kunzite can be used to foster self-discipline and a sense of responsibility

Aquarians tend to plough headlong into situations without thinking – brown zircon can encourage them to look before they leap

are concerned. That's not to say that Aquarians are in any way perverted, it's simply that the water-bearer likes to experiment in all spheres of life, not just in the laboratory!

Their stones are brown zircon, also called malacon, and chiastolite. Most brown zircon, as the name suggests, is brown in colour. However there are different-coloured varieties, such as orange (also called hyacinth or jacinth) zircon, yellow zircon and clear (also called jargoon) zircon. In fact the name zircon is derived from Persian words meaning golden colour. With the exception of the colourless stones, all these natural hues are suitable here, most being translucent.

Chiastolite is a variety of andalusite and is a fascinating stone. In cross-section, a four-armed, dull gold and faintly glittering cross can be seen. The stone itself is quite dark in colour but close study of the stone reveals the detail and beauty of the cross.

Many zircon crystals contain traces of uranium and thorium and the alpha particles emitted by these radioactive elements damage the crystal lattice. They are sometimes heat-treated to restore the crystalline structure, a process that has been practised since ancient times and turns them blue or yellow to create attractive gem stones.

The fact that these stones are radioactive and that they cannot attain gem-quality without being heat-treated make them highly appropriate for the inventive sign of Aquarius. Appropriately, its ruling planet Uranus governs all forms of energy, especially electricity and radiation. Keep several stones together as a reminder of the

potential for metamorphosis within us all. Aquarians tend to plough headlong into situations without really thinking through the consequences – brown zircon can encourage them to look before they leap.

Chiastolite has often been carried as an amulet. The cross within the circle is the astronomical symbol for earth, and as such, these cross-bearing stones may be used as a talismanic reminder of the debt we owe our mother planet. Chiastolite is also apposite here because of its individuality. Each stone is different, with endless variations of the cross and, used as a focus for meditation, may help Aquarians to acknowledge the responsibilities that result from their innovative ideas.

Moon stone
OPTICAL CALCITE
Those born with the moon in Aquarius possess ingenuity and tend to be progressive, while veering towards inefficiency, frivolity

and erratic behaviour. Aquarians may also have a profound understanding of the necessity for global improvement and the importance of happiness and equality for all. This is where the idea of Aquarius as the sign of 'universal brotherhood' comes from.

The stone for these Aquarians is optical calcite, also called iceland spar. This colourless, transparent form of calcite demonstrates double refraction, a property shared less obviously by many other crystals. By splitting light into two rays, any object viewed through the mineral is seen as two images. To observe this, draw a black mark or dot on a sheet of white paper and place a clear piece of optical calcite on top of it. Slowly rotate the stone, keeping it flat against the paper, and watch as a spectral image revolves around the real one. The thicker the specimen, the greater the spread of the images. The stone's ability to 'double' objects seen through it has given it the reputation of doubling magical energy in rituals.

Because Aquarians tend to see issues very

Optical calcite can shed light on dilemmas and difficulties, especially when it's impossible to decide what direction to take

much in terms of black and white, during meditation, optical calcite can help them to see the middle ground – something they often badly need! Talismanically, optical calcite can shed light on dilemmas, especially when it's impossible to decide what direction to take because they all look equally good or bad.

Pisces

Key word
COMPASSION

Characteristics
SUBTLE, COMPLEX,
VERSATILE,
IMMODERATE

Ruling planet
NEPTUNE

Ruling stone
BLACK OPAL

Sun stone
TIGERIRON

Moon stone
WHITE ARAGONITE

Black Opal

PISCES (the fish) is ruled by Neptune, which was actually discovered in 1846, although its probable existence was well-known long before this date. The discovery of the planet coincided with a resurgence of interest in the paranormal, which has been associated with Neptune ever since. Neptune was god of the ocean depths, a mysterious and profound character who had changeable moods but nevertheless provided a bountiful harvest.

Pisces is the sign of understanding, the profound, universal understanding that transcends age, culture and experience. Pisces encompasses a little of all the other signs, which is why it is often portrayed as lacking its own character. In fact, Pisces is marked by its unique tendency to switch between extremes: the fish has great difficulty in moderating its own behaviour and emotions. At their best, Pisceans are empathic, imaginative, intuitive, artistic and creative, yet they can also be decadent, indecisive, confused and emotionally unstable.

Ruling stone
BLACK OPAL

Blue-green stones are symbolic of Neptune, and black opal (which despite its name is not black) is Pisces' foremost stone. This beautiful irridescent gem symbolizes the mysterious depths of the dreaming unconscious which is the particular domain of Pisces. Black opal can be used in any situation involving the paranormal, religion, spiritual leaders or philosophers, or where Piscean qualities are needed. It can also be used both to promote Pisceans' positive traits and to minimize their less desirable qualities.

In the Orient, opal represents truth and brings good fortune to honest and unselfish folk. Above all, it is a token of hope for a better future. Furthermore, it is reputedly empowered with the gift of true prophesy and, by extension, can assist not only in the revelation of past lives but also the divination of what may be hidden or lost.

Black opal only became available in the 20th century and was found first at Lightning Ridge, Australia. Pisces' ideal stone will contain both green and blue irridescence. The opalescent colour-play is caused by the scattering of light by minute spaces between the spheres of silica which give opal its structure. However, tiny cracks form easily and can weaken the stone, a 'defect' that once gained the stone an unlucky reputation. Water, absorbed through these cracks, enhances the light-play, grease however, will dull it, so take care when handling your specimens.

Pisceans are often seen as shallow because they take other people's opinions on board almost indiscriminately. The result is that they become less truly themselves in order to reflect a persona that they think others want to see. Black opal encourages Pisceans to explore their own depths and to be true to

Black opal encourages Pisceans to explore their own depths and to be true to themselves – not to become what others expect them to be

themselves – not to become what others expect them to be. Black opals are best worn as jewellery – but take care not to drop them as they are fragile and will fracture easily.

Sun stone
TIGERIRON

Those born under the sun in Pisces can be empathic, versatile and possess a vivid imagination. They may also be unworldly, unstable and have a tendency towards hypochondria. Pisceans can suffer from a confused mass of conflicting emotions, thoughts, ideas and beliefs – sometimes even their dress-sense is chaotic! When faced with crises they have a tendency to run away, either literally, metaphorically into their own dream worlds, or physically into drugs or alcohol as a means of dulling the mind and their empathic abilities.

Tigeriron can act as an anchor for Pisces, focusing the fish's sparkling mind on the realities of daily life

Others often see Pisceans as targets for abuse or bullying, as nonentities to be ignored, or as potential worshippers – the blind devotion Pisceans display in their romantic lives is legendary. You should remember, however, that Pisceans also contain a little of all the other signs, so caution is advised: push the fish too far and you risk a positively Taurean stubborn refusal to budge or an outburst that would do a furious Scorpio proud! And in the end, the Piscean will probably disappear, leaving you with a strange emptiness in your life. The fundamental stone for Pisceans is tigeriron, a naturally occurring mixture of tiger's eye, red jasper and hematite. Often banded, it can resemble petrified wood but is more richly and distinctly coloured, sometimes even faintly sparkling.

Tigeriron can readily be worn as jewellery, in rings, bracelets or earrings. This intriguing stone, dark but with such a beautifully fascinating appearance, can act as an anchor for Pisces, focusing the fish's sparkling mind on the realities of daily life. Alternatively, small 'worry' eggs can be carried at all times to remind Pisceans of the need to balance their dreams and the 'real' world.

Moon stone
ARAGONITE

Those born with the moon in Pisces have the ability to articulate their imaginative visions while often suffering from typical Piscean self-doubt and uncertainty. Pisceans are naturally artistic in any number of ways: there are the Piscean artists, poets and dancers, the adventurous and creative cooks and interior designers, the mothers who weave the most beautiful and original stories at their children's bedsides. Pisces is never truly happy unless it can express, in some way, the visions it sees in its own mind.

Unfortunately, Pisces is also the least 'human' sign of the zodiac, and the fish often have difficulty integrating themselves with society. This can lead to them being viewed as 'weird' or even 'deranged' and while Pisceans may be fully aware of the effect they have on others, and why, they may find it almost impossible to improve the situation.

Their fundamental stone, white aragonite, may prove beneficial. Aragonite is usually white or colourless and is a polymorph of calcite, occurring in both sedimentary and metamorphic rock. It often has needle-shaped crystals, in large crystalline masses but its other forms are equally suitable for Pisceans.

Many marine organisms' skeletons are aragonite, and it also occurs as mother-of-pearl or nacre on the inside of shells as well as forming pearls. Its strong association with the sea makes it an appropriate talisman for Pisces. Aragonite's solidity can act to counterbalance Pisces' occasional instability of thought and emotion. It is best kept as an ornament somewhere prominent.

Aragonite's solidity can act to counterbalance Pisceans occasional instability of thought and emotion

Aries

PROBLEM	SOLUTION
Weakness, indecision, lack of strength	**Ruby** Wear ruby to promote stamina, forcefulness and determination.
Inability to concentrate, impatience, a tendency to exaggerate	**Rock crystal wand** Used as a talisman, the rock crystal wand will encourage Ariens to project their energies into specific projects rather than squandering them in all directions. It promotes mental alertness, adaptability and the ability to explain ideas clearly.
Nightmares, insomnia	**Imperial topaz** Keep this stone close to the bed or under the pillow to alleviate nightmares and promote sound sleep.
Emotional stress	**White marble** The protective qualities of this stone will help guard against and relieve emotional trauma.
Rashness in financial and legal matters	**Kyanite** (also called disthene). Kyanite's bladed shape can help channel Ariens' zeal into business and education. It is popularly used to aid understanding of financial and legal matters and promotes confidence and business acumen.
A tendency to be headstrong, incautious and impulsive	**Black flint** Use this opaque form of chalcedony to promote patience, a sense of duty and caution.
Rebelliousness, recklessness	**Purple fluorite** Handling this stone can impart a sense of natural order and help to structure thoughts. Turning a crystal from facet to facet while going over an idea in your mind can promote lateral thinking and spark new ideas.
Becoming involved in scandal, or being frightened of change	**Spinel** Use blue-green shades of spinel, which can form perfect octahedrons. It is used to guard against notoriety and scandal, and to induce responsiveness to new ideas.
Fanaticism, aggressiveness	**Sardonyx** Since Roman times, this translucent variety of chalcedony has been used to take a stand against oppression and injustice. The banding on the stones is allied to binding, restraining and limiting passionate urges, both in others and in oneself.

Taurus

PROBLEM	SOLUTION
Inflexibility, resistance to change	**Quartz cluster** Place the cluster somewhere prominent to promote diplomacy and pragmatism. Looking at crystals in a group or cluster, rather than as individual wands, enables you to observe relationships, as if in a family. This is ideal for occasionally autocratic Taureans!
Emotional dullness, romantic apathy	**Emerald** Wear emeralds as jewellery or carry them as talismans to aid expression of love and foster harmonious relationships.
Lack of imagination, tediousness	**Amber** This stone can help Taureans feel more creative, urging a more inquisitive approach in order to get the very most out of life.
Sluggishness	**Desert rose** Keep this fragile stone close by to enliven mind, body and spirit.
Obstinacy, stolidness, a tendency to create obstacles	**Garnet** (and any of the red-coloured garnet group of minerals – pyrope, rhodolite and spessartine). Garnets have a reputation for exciting interest, dispelling depression, strengthening the heart and improving blood circulation. They are best worn as jewellery to encourage tenacity and guard against inflexibility.
Extravagance, self-indulgence, poor memory	**Azurite nodules** Although azurite can form transparent, glassy crystals, it also forms granular, dull and opaque nodules that perfectly express the earthy Taurean nature. As a healing stone, it has a reputation for enlivening psychic abilities, especially in divination. It promotes charitable actions, good memory and self-understanding.
Workaholic tendencies	**Galena** This heavy mineral contains 86 per cent lead and a small amount of silver. Like the Taurean nature, it is very protective. It promotes self-discipline, frugality and management ability. Lead is also toxic, so galena must be handled with care. Anyone wishing to use galena as a talisman should place it in a glass-fronted cabinet where it can be seen but not accidentally touched.

Gemini

Cancer

PROBLEM	SOLUTION
Stress, tiredness	**Double terminator rock crystal** In stressful situations the double terminator can act like a safety-valve, deflecting stress. Be warned though, that the stone does not have a calming effect; rather it appears to strengthen the system so that you can continue under high pressure, so it is best used for short periods only.
Impatience, restlessness	**Citrine** This stone is thought to promote psychic awareness and, when used as a meditation stone, helps to clarify thoughts and ideas.
Inability to express or tackle intense emotions	**Apophyllite** can help Geminis feel more comfortable about developing their self-understanding, showing those capricious, inconstant people that emotion can be a beautiful thing!
Fickleness, aloofness	**Moldavite** This glass-like, non-crystalline mineral has a grounding effect, acting to remind Geminis that while their ideas and theories may be wonderful, they won't serve much purpose unless put into practice! It also promotes good humour and an affectionate response to friends.
Lack of self-assertiveness, a tendency to tire easily	**Fire opal** This translucent gem can appear to contain red and orange flames when light plays upon it and has a reputation as a passionate stone. It has an enlivening and refining effect, making it appropriate for Geminis who are often lacking in emotional intensity.
Superficiality	**Dumorturite** The whole stone can be faintly sparkling and, used as a meditation or 'worry' stone, shows Geminis that surface gloss isn't everything: sometimes it's necessary to look more deeply to find beauty and value.
Irrationality, confusion	**Stibnite** Dark silver or grey-hued specimens act to concentrate Gemini's mercurial mind on more rational, practical paths. However, stibnite is toxic, so handle with care. The long, straight crystals encourage Geminis to think logically . They make ideal ornaments which, kept at the workplace, will help focus the mind.

PROBLEM	SOLUTION
Overeating, other eating disorders	**Moonstone** Because of its association with the moon and the moon's cycle, moonstone is thought to help regulate eating habits so that they correspond to a more natural ebb and flow, balancing a tendency to overindulge and encouraging a healthier appetite in those people who eat too little.
A run of bad luck	**Silver** Popular as a talisman for attracting good luck and warding off misfortune, silver may be worn as an overall protection against the trials of life.
Muddled thoughts, illogicality	**Tumble-polished rock crystal** The best specimens for Cancer give the impression of a solid droplet of water. They should be used as meditation stones, allowing the mind a clearer insight into its own processes.
Emotional insecurity, 'clinginess'	**Amazonite** also called amazonstone. This pastel green or blue-green stone is opaque and faintly mottled with lighter hues. When turned in bright light, it radiates a silky, faint sparkle. It is a very 'earthy', worldly stone, representing the Cancerian's considerable strength of character, and is useful for grounding and stabilizing all the emotions.
Moodiness, frustration, stress-related illnesses	**Rose quartz** The shell-pink colour of this stone inspires comfort, relaxation and gentle warmth. Modern associations include sweetness, peace, contentment and joy. It is seen as an essentially feminine stone, particularly appropriate for nursing or new mothers, and useful for enhancing the maternal aspects of Cancerians.
Tension, changeability, fanciful ideas	**Blue fluorite** Worn as jewellery or used for meditation, blue fluorite can bring out the best of Cancer's considerable interpersonal skills! Fluorite has been used as an ornamental stone since classical times. It is reputed to quieten emotions and calm thoughts, making it ideal for meditation and highly appropriate for the occasionally overwrought Cancerian.

Leo

Virgo

PROBLEM	SOLUTION
A tendency to ignore other people's opinions	**Tiger's eye** This stone may be ideal for self-obsessed Leos, promoting a willingness to listen to others' points of view.
Arrogance, self-importance	**Rainbow aura** This is a manufactured item created by vaporizing platinum over the natural facets of a clear quartz crystal. This stone may be worn to remind Leos that they are not the only people in existence and to learn to take others into account!
Lack of self-confidence	**Gold** When worn, gold serves as a symbol of the status Leos feel they deserve, and so helps to reassure them of their own worth.
Confusion, bewilderment	**Diamond** The durability of diamond means it can help clear delirium and cut through muddled thinking, making it perfect for big-hearted but sometimes bewildered Leos!
Jealousy, vanity	**Malachite** Talismanically, solid, earthy malachite acts to calm the sometimes overpowering fiery nature of Leo in love.
Inability to concentrate	**Sunstone** This stone may be placed where the light can catch it to foster a vigorous approach to work. A small mobile hung in the window of the office would be ideal.
One-upmanship, dogmatism	**Sodalite** Crystals are rare: the translucent or opaque, blue granular form is more common and perfectly suitable here. For Leo, its solid, substantial appearance and smooth texture can have a calming effect on the occasionally overbearing Leonine nature.
Lack of direction	**Smoky quartz** Traditionally, this stone is used to lift depression, encourage pragmatism and boost enthusiasm. It can help Leos take control of their lives.
Over-exuberance	**Watermelon tourmaline** These transparent crystals are green on the outside and pink or red on the inside. The combination of colours and qualities can be beneficial, proving that active and passive qualities can co-exist peacefully – persuading Leos to tone down their exuberance, at least some of the time!

PROBLEM	SOLUTION
Stress, hyperactivity	**Rock crystal egg** The smooth, comforting shape of the 'worry egg' can help to ease stress and calm down an overactive mind.
Obstinacy	**Petrified wood** This aged stone can act as a reminder that too rigid a mental structure can result in a static, stagnant mind, much as a once-living, breathing, growing tree has been changed into stone.
Ill health	**Milky quartz** Traditionally, this stone promotes swift recovery and steady recuperation. You can either drink a milky quartz crystal essence remedy (see pages 70–73) or keep the stones within reach so that they may be frequently handled.
Frustration, frigidity	**Epidote** Also called pistacite because of its pistachio green colour, this is a hard, green and cream mottled stone that is particularly comforting to hold and handle. This is an excellent meditation stone and can promote delicacy in intimate relationships.
Tiredness, irritability, nervousness	**Rhodochrosite** also known as rosinca or 'Inca rose'. The rose-pink variety of this mineral is perfect here. On the one hand this stone is thought to increase physical strength and stamina; on the other, it can relax the body and bring a sense of forgiveness and love. Either property is fine for a Virgoan.
Anxiety, uncertainty	**Lace agate** This is a translucent blue variety of chalcedony. With its mottled pattern of wispy, lighter whorls, the stone has a reputation for calming and soothing both mind and body. Carry it with you and handle it when you feel pressurized to take any action that you may feel uncertain about.
A tendency to be obsessive about work	**Basalt** Although it can be dark grey, the pure black form is best used here. A basalt letter opener, for example, or other useful implement that can be kept in the workplace is most useful here. Touched or handled frequently, it can encourage a sensible attitude to work.

Libra

PROBLEM	SOLUTION
Depression, moodiness	**Chrysoprase** Chrysoprase can focus the mind on the physical aspects of life and promote enjoyment of the simple things. It is a 'feel-good' stone – touching it can promote feelings of happiness and of being at one with the world.
Distraction, a tendency to be unrealistic	**Pyrites** This stone helps Librans who are prone to being easily distracted from the matter at hand. Its heavy solidity can help bring them back down to earth.
Indecisiveness, lack of conviction	**Faceted quartz** There are many shapes of faceted quartz available and all are suitable here, although a square or baguette cut is best. The stone is stripped of its natural, irregular faces by careful polishing so that its shape is 'perfect'. This makes it ideal for perfection-seeking Librans, helping them to be analytical and persuasive.
Lack of confidence or courage	**Pink tourmaline** This transparent stone has a reputation for attracting kind thoughts and is useful in winning affection, trust and prestige. The redder varieties tend also to inspire confidence, bravery and honour.
A tendency to be easily swayed by others' opinions	**Blue agate** This is a beautiful translucent stone, varying in colour from bright pallid blue to the soft warm blue of early twilight. It is an immensely comforting stone to handle and should help Librans in their struggle for balance.
Inability to cope with adversity	**Black onyx** This is a soothing stone for Librans, reminding them that although true balance involves facing the darker side of life as well as the sunnier aspects, they have the strength to deal with difficult situations.
Over-sensitivity to criticism	**Malachite and azurite** Although azurite is rarer than malachite, they are often found together. Azurite alters to become malachite by reacting with common elements during weathering. A figure carved from this stone may be kept on view to remind Librans of their own worth and the validity of their own ideas.

Scorpio

PROBLEM	SOLUTION
Stress, anxiety	**Serpentine** Handle serpentine 'worry eggs' – their hard, cool smoothness and comforting weight will help to relieve stress. Meditating on the stone's patterns and colours can also open up the mind to new experiences.
Feeling miserable, pessimistic	**Sulphur** The sunny colour and unique characteristics of the stone show Scorpios that beauty can be found in the strangest of places! Sulphur should be handled as little as possible, but may be kept on show in a display case, where it can act to bring cheer to its environment.
Lack of self-control or self-confidence	**White opal** This is an ideal talismanic stone for Scorpios who are prone to angry outbursts, or who do not value themselves highly enough.
Prejudice, intolerance	**Rainbow quartz** Also called iris quartz, this stone can be distinguished from other natural forms of clear quartz by its internal flaws, which refract light in subtle yet spectacular ways. The clarity of the stone can focus the mind and encourage perception, while the lovely internal rainbows can persuade Scorpio to look for beauty, even where there appears to be none.
Jealousy, inner conflict	**Bloodstone** This opaque stone, also called heliotrope, is a green variety of chalcedony, speckled with red spots of iron oxide which gave rise to its common name. The Egyptians believed that it could help free them from bonds or shackles, demolish barriers and open doors, allowing them to escape their enemies. This makes it perfect for Scorpios, helping them to 'open the door' to their own pysches.
Restlessness, an inability to relax	**Red aventurine** This translucent mineral is usually green, often banded with tiny spots of colour and spangled with glittery mica flakes. Traditionally an all-round healing stone, it can also be used to promote insight, keen perception and understanding, which can help Scorpios in exploring their inner selves. The red variety is associated with relieving headaches, high blood pressure and depression.

Sagittarius

Capricorn

PROBLEM	SOLUTION
Materialism, boastfulness, dogmatism	**Sapphire** This precious gem can help balance these typically Sagittarian traits. It is also said to counter both physical and spiritual blindness.
Sluggishness, lack of energy	**Rutilated quartz** This stone is thought to boost energy, youthfulness and health and even to promote all those bodily defences that stave off the assault of advancing years.
Indecision, confusion	**White calcite** The clarity of this stone may help Sagittarians by acting as a focus for logical thought and decisive action.
Lack of co-ordination or staying power	**Window quartz** Windows are found near the tips of some quartz wands (see page 15). Depending on whether the window leans to the left or right, it is said to be either left-handed or right-handed. Right-handed windows are thought to stimulate the emotions and intuition, left-handed windows are associated with the intellect and logic. Occasionally, crystals have both, and these are symbolic of uniting the two halves of the brain and enabling tremendous creative potential – perfect for aspiring Sagittarians.
Unwillingness to participate in new adventures, fecklessness	**Prehnite, wavellite** Prehnite is a superb meditation stone that should keep even Sagittarians' restless minds occupied! Ideally, carry (or wear) a polished stone or keep a natural stone by the bed or on the desk. Wavellite symbolizes the restless nature of Sagittarius, and may be kept as a talisman of journeys taken or new journeys to come.
Impulsive behaviour, tactlessness	**Carnelian** The name is derived from the Latin word meaning flesh and thus the stone is associated with healing wounds as well as with stimulating carnal appetites. Sagittarians may wear or handle the stone to give added impetus to their search for wisdom!
Difficulty in solving problems or overcoming obstacles	**Obsidian** There are many varieties of dark obsidian, including 'Apache tear', which is particularly well placed here. Talismanically, obsidian is used to help overcome the many and varied obstacles that life throws in our paths at all turns!

PROBLEM	SOLUTION
Weakness, irresolution	**Jet** This stone is best worn as jewellery to promote strength of character and determination in the face of adversity.
Inability to cope with others' demands	**Yellow jasper** Capricorns are often seen as utterly dependable and so can often feel pressurized when others are leaning on them or making demands on their time. Carrying tumble-polished jasper can help to ease the strain.
Seriousness, lack of humour	**White chalcedony** Capricorns tend to take themselves very seriously. White chalcedony's bright, smooth whiteness encourages them to see the lighter side of life.
Inflexibility, bigotry, pedantry	**Rock crystal geode** All hollow geodes are suitable here and although they may look unpromising from the outside, they each contain a masterpiece of beauty within. They symbolize breakthroughs, so can help Capricorns break free from their conservative stolidity.
Self-centredness	**Jade** Nephrite, a form of jade, can be white, brown, purple, or most famously, green, which can be mottled with black. Also translucent jadeite is most often found in shades of green, or white with green spots. Both stones may be worn as jewellery, but even a small carved piece can help with issues of constancy, protectiveness and nurturing.
Prurience, inconstancy	**Red jasper** This opaque form of chalcedony is usually bright red due to the presence of iron oxide. Specimens are often veined or marked with dark, even black, streaks. Carry a tumble-polished piece of red jasper and handle it whenever stamina, persistence and strength of will are required!
Slapdash attitudes, lack of conscience	**Eilat stone** This is an opaque, blue-green mineral that is a combination of chrysocolla, turquoise and malachite and can be mounted and worn in a ring, so that its energy is ever-present. It promotes professionalism and a clear conscience.

Aquarius

PROBLEM	SOLUTION
Lack of self-discipline	**Kunzite** This lilac or purple transparent gem can be used to foster self-discipline and a sense of responsibility in Aquarians.
Recklessness, impulsiveness	**Brown zircon** For those Aquarians who tend to plough headlong into situations without really thinking through the consequences, brown zircon can promote caution and encourage them to think matters through before taking action.
Difficulty in solving problems, unwillingness to compromise	**Optical calcite** This stone can be very helpful during meditation to help Aquarians see the middle ground – something they often badly need because often their views are very black and white. Optical calcite can shed light on dilemmas and difficulties.
Lack of sympathy for others' ideas, aggressive bluntness	**Quartz sphere** This is the true crystal ball – but beware of glass imitations! A quartz sphere can promote an understanding of obscure ideas, and also objectivity, sensitivity and sympathy for other people's thoughts and views.
Shyness	**Aventurine** This translucent mineral is regarded as an all-round healing stone. It has the reputation of promoting insight, keen perception and heightened awareness. It may also bring luck in games of chance, presumably by awakening or enhancing precognitive abilities! It is best used as a 'worry' stone or for meditation to boost spontaneity and freedom of expression.
Clinical, cold attitude to others	**Rhodonite** The cheerful appearance of this stone is held to promote balance, engender a sense of proportion, a sense of humour and humanitarian feelings. Carry a tumble-polished stone and handle it whenever you feel your love for your fellow human-beings waning!
Unreliability	**Iolite** Also called water sapphire or dichroite, this is a gem-quality variety of cordierite. When viewed from the side, many iolite crystals appear yellow, while from the base they are a deep blue. Carried or handled during study, or used as a focus for meditation, this appealing crystal promotes co-operation, respect for others and democratic values.

Pisces

PROBLEM	SOLUTION
Lack of self-confidence	**Black opal** This gem encourages Pisceans to explore their own depths and to be true to themselves rather than try to be what other people expect them to be. Black opals are best worn as jewellery – but take care not to drop them as they are very fragile.
Having unrealistic goals or dreams	**Tigeriron** This dark but fascinating stone can act as an anchor for Pisceans, focusing their sparkling minds on the realities of daily life.
Mood swings, feeling over-emotional	**Aragonite** The solidity of this stone can help counterbalance Piscean's occasional instability of thought and emotion. It is best kept as a small ornament, somewhere prominent but safe in the room in which you spend most time.
Prejudice, listlessness	**Rock crystal phantom** This is a rock crystal, usually a wand, within which the image of another crystal can be seen as if growing inside the parent. Rock crystal phantoms are excellent meditation stones, fascinating to examine and explore. Small stones can be carried and used as a reminder to listen to others and to promote receptivity and intelligence.
A tendency to be too submissive	**Dioptase** Once seen in its crystal form this beautiful rich green stone is hard to forget. Dioptase can be worn as jewellery, or you may prefer to keep a crystal cluster within reach on a desk or bedside table. It promotes tenderness and deep, selfless love while balancing the Piscean tendency to be overly submissive and conciliatory in manner.
Lack of sexual desire, discontentment	**Dragon's eye** Here the golden chatoyancy of the yellow tiger's eye is replaced by a beautiful burgundy. Try keeping a tumble-polished stone, or as large a sphere as you can afford, beside the bed to promote sensuality and playfulness. At the very least it should inspire you to new heights of passion!

CRYSTALS FOR PERSONAL POWER

Most of us have dreams, hopes and desires – and crystals can help us achieve our goals. Whether you want to unlock your inner potential, find love or enjoy material success, a crystal talisman can help you find fulfilment.

Find out all you need to know about choosing your talisman, cleansing it, and giving it additional powers. You can even learn how to use pendulum dowsing to get in touch with your subconscious mind. Whatever your aspirations, this chapter shows how crystals can enrich your life and help you find happiness and success.

Using crystals as talismans

Most people have a rather hazy idea of what an amulet or talisman looks like: the usual image is of a gold pendant encrusted with priceless precious stones and swathed with cryptic magic symbols straight out of the Arabian Nights – nice if you can afford it!

Yet literally anything can be talismanic, provided it means something to you. Even Peter Pan's 'happy thought', which enabled him to fly, is a talisman. Talismans can be likened to stepping stones, helping us to progress towards the fulfilment of our dearest dream. Using the stepping stone analogy, some are broad and secure while others demand a sure eye and nimble feet and offer no consolation to the blindly optimistic.

Crystals and other stones have been used for centuries to represent particular hopes and wishes. By choosing the specific stones that best reflect your own personal wishes and wearing them or carrying them around with you, you can tap into this traditional power that helps to make dreams come true.

We can all think of things we'd like to change about ourselves and the world in which we live. If you were handed a magical wand and offered the chance to make a few changes, how long would it be before you tried to make a few immediate improvements in, say, health, wealth, or love?

A bold necklace of blood red rubies makes a loud statement that the wearer is not only full of confidence but is looking for an exciting time. Delicate earrings of blue celestite, on the other hand, speak volumes about the wearer's gentle and contemplative character. We all make instinctive emotional responses in the face of such clear signals. Learning to use jewellery as a means of sending the right message is an art in itself, but crystals and other minerals are capable of much more subtle effects. Talismanic magicians have for millennia used the different elements of the natural world to affect their lives in an ostensibly supernatural way.

The power of crystals to attract luck and dispel misfortune is supported by a wealth of anecdotal evidence, but is not recognized by science. The ability of such minerals to influence the mind of the person who uses and wears them, however, is widely attested.

Self empowerment using crystals works at the same level as our human capacity for looking at a picture of a crying child and

being moved to tears ourselves, or of a glorious sunrise raising our hopes. Although the stones trigger our responses, we ourselves are the source of the magic.

WHICH STONES TO USE?

Think long and hard about what it is that you really want. You can even write your wish list on paper, so that you can review it whenever you need to. Then put your list to one side for a while. When you come back to it later, try to prioritize the areas in your life that you are most likely to be able to change and sideline (but do not discard) the least promising … Once you have decided on your goal, you will probably find that a certain sequence of events will lead you to your heart's desire. At this stage, any number of different talismans can be employed to help you towards each target while one particular stone can be used constantly, representing your ultimate goal.

In a way, choosing a talisman is like picking a friend! To find a particular talisman to help you achieve your heart's desire, all you need do is to refer to following pages: a suitable stone is suggested for more than 100 different scenarios and areas of your life that you might want to focus on.

This rock crystal, found in Brazil (left), is a mix of golden beryl, morganite, aquamarine and emerald. A rock crystal pendant is shown above.

How to choose your talisman

GOAL	STONE	GOAL	STONE
New beginnings: anything from moving into a new house, starting a new job, or having a baby.	Topaz	To help overcome self-indulgent tendencies. Keep in front of the mirror!	Azurite nodules
To help steady impulsive behaviour and create new opportunities in your life as a whole.	White marble	To create self-discipline at work – especially where you are prone to work so hard that your personal life suffers.	Galena (Toxic, handle with care.)
To focus and direct your mental energies towards a desired goal, especially where work is concerned. Try carrying this talisman when requesting a pay rise!	Rock crystal wand	To promote constructive thinking and mental resourcefulness.	Charoite
To calm down a tempestuous romance and encourage consideration in your partner.	Peridot	To promote and enhance artistic inspiration and appreciation. Can be useful in art and music tests and interviews.	Chrysocolla
To promote leadership. Try wearing it for success when engaging in any competitive activity.	Ruby	For common sense. Used as a worry stone, it can help you to see the funny side of a situation.	Unakite
To channel energy and enthusiasm into formal procedures. Keep at hand when trying to express your great new ideas on paper for examination and consideration by others. To help you understand legal or financial jargon.	Kyanite	To assist the development of circumspection in dealing with others, especially at work. To make speaking to those in authority a little easier.	Citrine
To assist concentration. Keep a piece nearby at your workplace to help you keep your mind on your work!	Black flint	To help acceptance of your own feelings. Useful for 'uptight' individuals, allowing them to relax and gently explore their fears and motivations.	Apophyllite
To develop your own new ideas and promote lateral thinking abilities. May also act to temper a sense of personal infallibility. You could carry one with you when driving!	Purple fluorite crystal	For situations where clear communication is necessary. You could carry it with you at all times.	Rock crystal double terminator
To encourage responsiveness to new ideas from others, and overcome obstructive single-mindedness.	Spinel	To encourage an affectionate response to others and a love of variety in relationships. Can be useful when embarking on a first date, or going out with people you only know slightly.	Moldavite
Use to temper aggressive urges. May be used where subtle bullying or harassment are present, to bolster courage to deal with, or speak out against, the situation.	Sardonyx	To develop keen powers of observation.	Fire opal
To promote stability, help combat feelings of insecurity and to attract treasures (especially antiques).	Amber	To promote ease of expression. It may also prove useful if you find conventional methods of education difficult to handle. Why not carry a tumble-polished piece to class to help keep your mind calm and clear.	Dumorturite
To encourage sensuality and enliven a slow metabolism. Keep in the bedroom and don't handle too often.	Desert rose	To assist in ordering chaotic thinking, and creating a logical approach to problem solving. You could keep under glass as a meditation stone.	Stibnite
Helps dispel resistance to change and promotes mental adaptability. Keep in the workplace.	Rock quartz cluster	To promote spontaneity and flashes of insight.	Ametrine
To help overcome stubbornness and self-created obstacles. Perhaps most beneficial when worn as earrings.	Garnet	To help overcome impractical tendencies.	Aqua aura
		For discovering hidden potential, especially mental or occult abilities. Use as a focus for meditation, and carry with you to help give your mind a rest from work during meal-breaks.	Spectrolite

To develop self-analysis or help you to break bad habits.	Onyx
For protection or to promote caution. Useful to keep on your desk, especially in a new job.	Aragonite
To help alleviate digestive or eating disorders, or to promote a calm and relaxing atmosphere. Useful for busy mothers.	Moonstone
For overall protection against the tribulations of life. It won't make them disappear, but may help you endure them with patience and a sense of humour!	Silver
As a meditation stone, to develop good memory and a quick understanding, especially of children and their problems.	Tumble-polished rock crystal
To promote maternal feelings and the ability to cherish others. Useful for the harassed executive!	Amazonite
To help alleviate stress-related illness, combat frustration (especially in the home) and a tendency to moodiness.	Rose quartz
To promote excellent interpersonal relations. You can wear this at work.	Blue moonstone
The ultimate protective stone, used for deflecting negative influences and encouraging tolerance. It may also help control panic attacks.	Specular hematite
To develop a well-balanced outlook on life, and free your creative abilities.	Strombolite
To encourage responsiveness to emotional stimuli, and promote self-confidence.	Smithsonite (copper bearing)
To improve tenacity and willpower, especially in personal relationships.	Turitella
To lessen your dependence on the approval and opinions of others.	Ammonite
Encouraging a deep understanding of your own inner resources, and developing the desire and willpower for success.	Tiger's eye
To promote altruism and purity of intent. Also an overall healing mineral, which fosters physical and emotional well being.	Gold
Embodies beauty, confidence and creativity and can be worn to promote these qualities.	Diamond
To help combat arrogance and encourage a willingness to learn.	Rainbow aura
To create faithfulness and physical energy – ideal to wear or keep (perhaps as a carved animal) in the bedroom!	Malachite

To promote courage and energy, and overcome ostentatious tendencies.	Sunstone
To encourage nobility of ideals and personal dignity, help banish the fear of matters beyond your control and help you deal with guilt.	Sodalite
To combat a refusal to see the world as it really is, warts and all. Develops a sense of realism and pragmatism.	Smoky quartz
To promote a dynamic personality while at the same time acting as a reminder that there is always someone who knows more, or can perform better, than you!	Purple fluorite
To encourage selfless love and altruistic tendencies. Useful in the home, especially if there are children about.	Watermelon tourmaline
To promote justifiable pride in your achievements. Can be very useful for those who have had their ideas and work denigrated over time, to the extent that they may feel as if they have nothing to offer.	Leopardskin jasper
When rational thinking and a logical approach to problem solving are required.	Petrified or opalized wood
To help speed recovery if you're feeling under par.	Milky quartz
For the successful resolution of projects. The shape of the stone makes it an excellent talisman for a good harvest, whether of ideas, individuals, or material.	Stilbite
As a means to focus your mind when contemplating the future, or as a worry stone if you feel insecure (especially at work).	Rock crystal egg
When you want to bring out the gentle side of intimate relationships. Keep it with you if matters tend to get out of hand, or if you would like your partner to be more considerate!	Epidote
To help overcome shyness or nervousness.	Rhodochrosite
To aid personal integrity. Can help you stand firm when under pressure to do something about which you feel uncertain or worried.	Lace agate
To promote efficiency and an appreciation of the holistic approach to life.	Black basalt
To combat overly critical or faultfinding tendencies, either in yourself or others.	Mauve opaline
When real precision or attention to detail are required. Promotes perfectionism and critical judgement.	Larimar

To encourage analytical skills; to help calm the nerves when you feel control slipping – especially if others are causing the problems.	Brecciated jasper
To promote eloquence and combat a tendency to be easily distracted. To help you keep your feet firmly on the ground!	Iron pyrites
To encourage honesty. Use in any situation where you suspect your companions are being less than truthful – or to give yourself the nerve to be honest yourself in uncomfortable circumstances.	Selenite
To combat indecisiveness. Good if you know you are going to be faced with difficult decisions, or even if you can't decide where to go for dinner!	Faceted rock crystal
To promote enjoyment of the simple, natural things in life, especially if you feel that your life is becoming too complicated.	Chrysoprase
When you want to be fair in your dealings with others – and want them to be fair to you.	Pink tourmaline
If you have a tendency to be too easily swayed by others' opinions, try using this stone to contemplate the worth of your own ideas and beliefs.	Blue agate
To encourage impartiality and diplomacy.	Black onyx
To help ease a situation in which you have to compromise in order to keep the peace or ensure matters run smoothly, whether you want to or not.	Sugilite
To help combat a tendency to be overly sensitive to the criticism of others. It is worth remembering that criticism is most often designed to be helpful, not malicious.	Malachite, azurite
To help overcome low self-esteem and help foster self-confidence.	Snowflake obsidian
For helping self-control and self-reliance. Beneficial for those who feel themselves at the mercy of other people for much of the time.	Golden beryl
To help bring your inner beauty to the surface, where it may be admired by others.	White opal
To enhance your perceptiveness – and your ability to keep secrets. Useful for those who have a tendency to gossip.	Rainbow quartz
To promote passion and a love of the dramatic. Ideal for the bedroom!	Bloodstone
To encourage self-reliance and resourcefulness.	Red aventurine

To help overcome manipulative tendencies, both your own and others.	Falcon's eye
When you need to guard against loss of emotional control, especially in intimate or hostile circumstances.	Black calcite
To guard against ruthlessness. Occasionally, this stone contains crystals of pink tourmaline, which may be particularly useful in enabling lovers to share deep feelings, hopes and fears, thereby strengthening their relationship.	Lepidolite
To promote investigation into the occult and paranormal, while helping to protect against cravings for new and potentially dangerous sensations.	Hemimorphite
To promote inner peace, strength, and an appreciation of the need for conservation of resources, both your own and those of the planet.	Serpentine
To stay young at heart and bursting with vitality!	Rutilated quartz
To promote independence and an inquiring mind. Excellent for anyone going away from home to engage in a course of study or training.	Clear calcite
To encourage directness of speech, thought and action, and assist co-ordination, both physical and in the workplace.	'Window' quartz
To encourage a willingness to try new experiences. Can be useful for those about to go on an adventure holiday!	Prehnite
To combat lack of tact and dangerously impulsive behaviour. Can be useful to take with you when travelling.	Carnelian
Embodying benevolence, optimism and learned wisdom, this stone may be worn to engender and enhance these qualities.	Sapphire
To help combat depression.	Obsidian (Apache tear)
To promote prophetic dreams and visions.	Tanzanite
Symbolizing spirituality and strong moral codes, you can wear this stone to strengthen your resolve when faced with temptation.	Turquoise
To restore your faith in the human race and also help curb an extravagant streak.	Lapis lazuli
To promote dependability and economy. Can be useful to carry when shopping – it may help you avoid spending too much!	Yellow jasper

To help temper materialistic and mercenary qualities. Can be useful to keep at work if you find the atmosphere a little unfriendly.	White chalcedony
When mental clarity is required or to curb pedantic tendencies. May prove beneficial when taking tests or exams.	Rock crystal geode
To promote constancy in love and engender protective tendencies – an ideal gift for a loved one or for new parents.	Jade
To enhance sexual stamina! Keep by the bed to liven up your romance.	Red jasper
To help curb selfishness and promote ambition for others – especially children.	Natural azurite
For endurance and determination and to help you overcome obstacles in your life.	Jet
To encourage shrewdness and acquisitive tendencies. Can be useful for entrepreneurs and those starting their own business.	Violet spinel
To help enhance professionalism in your chosen career.	Eilat stone
To promote prosperity and growth in all things, this general luck-bringer may be used in both business and your private life.	Moss agate
To remind you of the possibility of change and the potential for human development that is within us all.	Brown zircon
To help shed light on problems and dilemmas, thus making them easier to solve.	Optical calcite
To create objectivity and clear-mindedness. Anyone involved in education can benefit from keeping this stone with them: it doesn't need to be large; a tiny sphere less than the size of a marble is easily portable and very appropriate.	Rock quartz sphere
For those wishing to break out of a rut, to help promote freedom – of thought, of speech and lifestyle.	Aventurine
To encourage humanitarianism. Useful if you wish to counter bigotry or prejudice.	Rhodonite
To encourage respect for the ideas of others and encourage a democratic outlook. Used as a focus for meditation, this stone is a gentle reminder that life is not black and white, and that beauty itself can be a passport to illumination.	Iolite
To encourage originality and also to prevent wasting resources through impractical ideas.	Black tourmaline

To promote self-discipline and a sense of responsibility. Useful to have with you when looking after other people or their property.	Kunzite
For those with a tendency to rely on stimulants (coffee, tobacco, alcohol) to keep going when they really should rest, this stone may be effective in enabling them to relax naturally.	Aquamarine
To encourage the ability to absorb information. Try keeping a worry egg or smooth piece of the stone at your desk or workstation where it can be frequently handled.	Orthoceras
To counter impracticality and chaotic thoughts, this can act as a 'grounding' stone for people who have a tendency to unworldliness.	Tigeriron
To help create the ability to give outward expression to dreams and visions. Ideally, keep a piece beside the bed, sitting on top of a notepad and pen, and try to write down your dreams as soon as you wake.	White aragonite
To enhance receptivity and generate a sense of humour.	Rock crystal phantom
To promote devotion and selfless love. This is an ideal stone to be taken to a refuge or used as a focus of meditation during a period of retreat, as it can enable you to commune with your inner self.	Dioptase
To inspire novelty and playfulness in sexual matters! Keep some by the bed.	Dragon's eye
To promote refinement, compassion and good taste, or to help you maintain control in hostile or stressful situations.	Celestite
To encourage solitude and a feeling of self-containment, even in the midst of a busy office. Also to help expand the mind, inducing a wider perspective that can be both inspiring and a little frightening.	Black tektite
To help enhance intuition and imagination, or to bring out artistic talent.	Amethyst
Embodying the power of dreams, mysteries and the occult, this stone may help anyone wishing to enhance their understanding of – or simply explore – these things.	Black opal
To help promote self-awareness and curb tendencies towards morbid sentimentality.	Tourmalinated quartz

How talismans can help

Talismans can be regarded as keys. They are capable of opening up a future which had seemed closed to you. Such is the traditional promise, but if they are such powerful magical tools – why isn't everyone using them? The answer is that everyone is. Many of us have trinkets or clothing that we regard as lucky and from that favourite shirt to the car you drive, everything has its own talismanic power and purpose.

Of course, you can't wear that lucky shirt every day – whereas crystal talismans can accompany you anywhere. Carry them in a pocket or purse, or wear them as jewellery: earrings, rings, pendants or bracelets, the list is practically endless.

The talisman's power radiates like an aura, simultaneously protecting you from misfortune and failure and stimulating the optimum conditions to speed you on to your desired goal. Simply by twirling a stone between your fingers you can invoke its power immediately, whenever it is needed. Even when you are unconsciously toying with the stone you can reinforce the wish and quickly draw out its power to help.

Two areas in which many of us experience problems are love and education, so to get into the swing of using a talisman, here are a few suggestions to help improve love and learning.

TOP TEN LOVE STONES

Do you want to strengthen an existing relationship or get to know a passing acquaintance better?

Emerald will help you to build and develop a rapport.

Are you faced with having to choose between potential (or current) lovers?

Moldavite can help you make the right decision.

Are you so infatuated by someone that you can't think straight?

Amazonite will help you to cool down, get your feet back on the ground and think more clearly.

Has your relationship become rather staid and boring or is it going nowhere?

Choose **malachite** to give it a new dimension.

Are you worried about moving in with a partner, or settling down?

Epidote will help you understand what living together could actually mean, emotionally, financially and practically.

Do you feel that the balance of power in your relationship is uneven?

Chrysoprase can be used to promote harmony and to help both partners relax and enjoy each other for who they are.

Are you unconvinced by the power of love?

Bloodstone will show you that love can move mountains when both partners are dedicated to achieving the same goal.

Are outside pressures taking their toll on your relationship?

Prehnite or **wavellite** will help a couple to overcome worries and pressures that seem beyond their control.

Are you unfulfilled or stuck in an unhappy partnership?

Aventurine can assist you in breaking away from an unloving or destructive relationship.

Are you investing so much time and energy in your relationship that you don't have a life outside it?

Dioptase can help you to find a passion for an ideal, a good cause or even a faith, outside the bounds of a stable relationship.

TOP TEN LEARNING STONES

Are you nervous or worried about the first day of your educational course?

Carry **black flint** on the first day to help you break the ice with your new colleagues and fellow students.

Do you think you might find it difficult to settle into the college or class?

On the second day, take **basalt** with you to help you develop a routine.

Finding it difficult to take in so many new facts and figures?

Once the course is under way, **black onyx** is perfect to help you assimilate and weigh up large quantities of information.

Have you been invited to attend a social event or party during the course?

Black tektite could give you the chance to project other aspects of your personality and show you're more than just a model student!

How can you put all those new ideas and information to best use?

Pick **fire opal** to inspire and motivate you.

Is your course accountancy-based?

Lace agate may help you keep long-term goals clearly in mind.

Is your course an artistic one, such as creative writing?

Amethyst is your ideal talisman.

Do you tend to put off doing course homework, or fail to do as much as you should?

Kunzite will help to promote self-discipline and a more responsible attitude.

Are you feeling nervous about travelling away from home to study or undertake a training course?

Take some **clear calcite** with you to promote independence and an inquiring mind.

Do you find it difficult to concentrate?

Make use of a **rock crystal wand** to help you focus and direct your mental energies into your work.

Purifying your talismans

Talismans quickly become charged with memories. They have the potential to become as precious to you as a favourite souvenir, like the pebble that takes you back to a wonderful beach holiday. However, just as stones are believed to be storehouses of healing energy or talismanic power, they can also soak up impressions from their environment, some of which may be undesirable. It is a good idea to rid the stones of these unwanted elements by cleansing or purifying them.

Geode

You may use earth, fire, air or water to purify your stones, provided the stone can withstand the treatment. If the stone is fragile or of uncertain provenance, you may also decontaminate it quickly by resting it for a while in a rock crystal cluster.

To purify your stone with earth, cover the stone with non-synthetic cloth and wrap it well. Bury it in the ground, no deeper than the distance from wrist to fingertip. Leave it for three nights and retrieve it on the third day.

If using water, immerse the stone in fresh still water, changing the water three times a day for a full three days. To purify your stone with air, lightly toss the stone through plumes of incense smoke. Repeat this a total of nine times.

Finally, if using fire, allow the stone to bask in sunshine for one hour. Alternatively, place it within a circle of three white candles and allow it to sit in their flickering light for one hour. (Just remember never to leave burning candles unattended!)

You may wish to recite a few appropriate words. For instance, at the commencement of the purification, try the following: 'Welcome [name of mineral], in the hope of strength, truth and beauty, I offer you purification by the element of [earth/water/air/fire].'

After the allotted time, you may add: 'Purified by [earth/water/air/fire] in the trust of strength, truth and beauty, I treasure you [name of mineral].'

THE CLEANSING POWER OF SALT

Salt, also known as halite, is often regarded as a magical catalyst. If you are planning an elaborate rite of talismanic magic, you can use salt to increase the power of the cleansing and banish any residual traces of inharmonious character that may cling to a stone from past use or abuse.

To do this, you need to make your own halite crystals. Simply dissolve a few tablespoons of pure rock salt in a quarter of a

Rock crystal can be held like a magic wand, and used to project your willpower towards your goal

pint (140 ml) of still mineral water, warmed if required to speed up dissolution. While the salt is dissolving, you may like to focus your mind on a similar mental or spiritual transmutation. Here, salt – a solid – is being transmuted into a 'higher' or more subtle element – liquid. At the same time, you can imagine the dissolution of the ego into the realms of unified spirituality.

Pour the saline liquid into a bowl or tumbler. If any of the salt has not dissolved, then strain the liquid carefully to prevent any of the original crystals going into the container. Cover the bowl with a tea towel or material to keep dust or impurities from getting into the liquid and put it in a warm, dry place, such as a mantelpiece or sunny windowsill.

As the water slowly evaporates, new crystals will start to form at the bottom of the bowl, eventually becoming a miniature landscape of tiny cubic crystals. These look fragile but are actually remarkably tough and will probably need to be carefully chipped out. Occasionally, a few individuals will be so well-formed that they deserve to be kept aside as specimens.

The remaining crystals can be crumbled into a heap of glittering cubes – building blocks of magical power. You can now meditate by considering how these crystals have grown out of the liquid, reflecting on how this phase of the great cycle of transmutation is the consolidation of spiritual energy.

The recrystalized halite, empowered by your meditations, may then be scattered in a ring around any stone that you wish to purify or charge with positive energy. Lighting a single white candle while you contemplate the stone within its protective circle will invariably be time well spent.

These shiny crystals can be kept for years and frequently re-used if stored in a clean, dry place. Use a pinch of these crystals to prepare a special cleansing bath, or dissolve a small crystal in a pint of water and sprinkle it wherever needed, such as a room in which the atmosphere needs to be cleansed.

Giving your talismans more power

To bring out a stone's innate qualities and to imbue your talisman with additional power that you can use to influence your life, you can make the stone the focus of a simple ritual. To infuse it with energy, the stone is surrounded by colours, metals and scents that naturally radiate the power you want to harness, which is then directed into the stone by the action of your will. This may sound complicated, but it is actually very straightforward if you take it step by step!

To choose a ritual that best suits your purpose, refer to the chart opposite. You will see that a stone is suggested for each scenario, together with the day on which you should perform your ritual, the metal you should bring with you and the scent with which to perfume your ritual – for example, if a pine scent is suggested, then you can either wear a pine-scented deodorant, or take a sprig of pine with you. Arm yourself with an item of the appropriate colour, or wear clothes in that colour, find the right place for your ritual and you're ready to begin!

Hold the stone and begin to concentrate on your desired purpose. Feel the strength of your will flowing into the stone: close your eyes (if safe) and feel an answering warmth flow back from the stone and into you. Imagine the stone becoming a friend, a trusted companion, a partner. If you like, imagine yourself in the situation with which the talisman is going to assist you. When you find your concentration slipping, it's time to stop. Gently breathe onto your stone to complete the cycle.

You can make your ritual as elaborate or simple as you like, incorporating elements other than those suggested (for example, you can use incense, candles and even music that summons up the essence of your desired result). You are creating your talisman, so make the ritual as intensely personal as you wish.

TEN SIMPLE RITUALS

• *To improve your career prospects*
Stone **Ruby**
Best day **Tuesday**
Metal **Iron (e.g. iron nail)**
Scent **Citrus**
Colour **Scarlet**
Where to do your ritual **A place associated with competition, such as a playing field**

• *To attract love and romance*
Stone **Emerald**
Best day **Friday**
Metal **Copper (e.g. copper bangle)**
Scent **Pine**
Colour **Deep green**
Where to do your ritual **Find a spot where there are plenty of roses, such as a rose arbour**

• *To bring more opportunities into your life*
Stone **Rock crystal double terminator**
Best day **Wednesday**
Metal **Tin**
Scent **Herbal**
Colour **Palest blue**
Where to do your ritual **Near a stream or river**

• *For personal balance and to relieve stress of all kinds*
Stone **Chrysoprase**
Best day **Friday**
Metal **Copper**
Scent **Fresh**
Colour **Grey-green**
Where to do your ritual **Anywhere that feels restful and harmonious**

• *To help you change and take control of your life*
Stone **Serpentine**
Best day **Sunday**
Metal **Steel**
Scent **Musk**
Colour **Red, black**
Where to do your ritual **Find an out-of-the-way, solitary place**

• *To bring luck in legal and financial matters*
Stone **Sapphire/sodalite**
Best day **Thursday**
Metal **Brass**
Scent **Woody**
Colour **Royal blue**
Where to do your ritual **Under the canopy of a mature tree, e.g. oak**

• *For calm and endurance when taking tests*
Stone **Jet**
Best day **Saturday**
Metal **Lead**
Scent **Smoky**
Colour **Black**
Where to do your ritual **Find somewhere quiet where you can concentrate easily**

• *To help you find like-minded friends and get on with colleagues*
Stone **Kunzite/amethyst**
Best day **Monday**
Metal **Titanium**
Scent **Sharp**
Colour **Bright purple**
Where to do your ritual **Try to find somewhere which makes you feel invigorated**

• *For health and wellbeing*
Stone **Rock crystal egg**
Best day **Wednesday**
Metal **Tin**
Scent **Floral**
Colour **Pale yellow**
Where to do your ritual **In a garden blooming with flowers**

• *To assist you in spiritual quests*
Stone **Black opal/spectrolite**
Best day **Any, but Thursday may be slightly more favourable**
Metal **Platinum**
Scent **Exotic**
Colour **Turquoise**
Where to do your ritual **Ideally by the sea, or somewhere that makes you think of the ocean**

Step-by-step ritual

If you have the time, you can use your talisman for a more involved personal ritual that will attune the stone to you and boost its power to help you achieve your goal.

Bringing in elements of the easy rituals already suggested, this step-by-step ritual takes a little longer and requires some concentrated effort – but the benefits are invariably worthwhile.

Do not start working with your new stones until you are happy that they are not carrying any distractions for you. If, for example, you keep thinking of its previous owner whenever you look at the stone, or you imagine some inexplicable event or sensation, you may like to cleanse the stone again (see page 144). When you feel comfortable with your stones you are ready to connect to the energy that they represent.

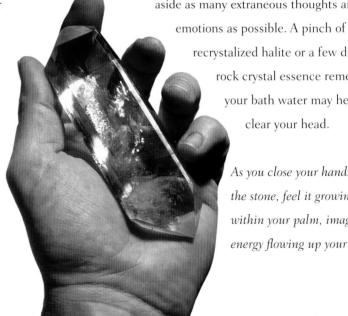

Aqua aura

Set aside anything from 30 minutes to an hour and a half for your ritual, choose a comfortable place where you will not be disturbed and where you can use a surface, such as a clear coffee table, dressing table, or even a space on the floor.

PREPARING FOR YOUR RITUAL

First, you should clean the site for the ritual by lightly sprinkling or misting it with a rock crystal essence remedy (see page 70–73) or water prepared with recrystalized halite (see page 145). Close the curtains and switch off as many lights as

you can without making the room too dark. Wrap the stone in cotton or put it in a box and place it on the table or floor in front of you, then surround it with items suggested by the 'Ten Simple Rituals' chart on page 147 to help focus the power you are working with.

Set up an odd number of candles in appropriate colours – if your stone is an emerald, for example, you could choose green candles (although white candles will do for any stone).

Organize any incense but leave it unlit. You can then cleanse yourself, and it is important not only to clean the body but also to put aside as many extraneous thoughts and emotions as possible. A pinch of recrystalized halite or a few drops of a rock crystal essence remedy in your bath water may help to clear your head.

As you close your hands around the stone, feel it growing warm within your palm, imagining its energy flowing up your arms.

STEP ONE

As you enter the place that you have cleansed and made ready, remember that you are seeking something, and what you receive will largely depend upon your actions – both in body and thought. It is important to concentrate on what you are doing and to resolutely ignore any trivial distractions.

Light the central candle (extinguish any other light source) and the incense. Then light any other candles. Allow your mind to clear. Close your eyes and soak up the atmosphere. When you open your eyes, allow them to fall where they will and watch as your mind follows the run of associations sparked by each ritual object.

When you have observed them all, close your eyes and imagine them all carefully – not one by one, but all together just as you arranged them, radiant and close to you. These are the fragments of an ancient and mystical puzzle that you must assemble in order to clearly see the power in your chosen stone. The more clearly you can divine this vision, the more powerful the talisman will be.

STEP TWO

When you are absorbed in contemplation (or have reached a point where further concentration is fruitless) open your eyes, reach for the stone and unveil it. Study it as if you had never seen

Rock crystal wand

Rose quartz

it before, all the while remembering the emotions, ideas and images that you have been so ardently pursuing. Close it in your hands as you close your eyes again. Try to feel the stone growing warm against your palms, imagine its energy flowing up your arms and gently cascading into your shoulders, its essence filling your entire body like a re-vitalizing fragrance.

STEP THREE

Once you are full from the tip of your toes to the top of your head, take a deep breath, open your eyes and carefully blow onto the stone in your cupped hands – imagining all the stone's essence inside you being borne out of your mouth and into the stone. Having been attuned to the stone, this return of its essence now symbolically attunes the stone to you.

Replace and rewrap the stone. Extinguish the candles in reverse order and tidy away all the other items, leaving the stone until last.

WORKING WITH OTHERS

Rather than offering to create a talisman for a friend, it is always better to help them perform their own ritual – otherwise, attuning the stone properly can be difficult.

However, if you cannot get together with your friend, or if the talisman is intended to help heal a pet, you can take a strand of their hair and position it between you and the stone as you exhale.

Three key rituals

Are you under pressure at work or home and suffering from stress as a result? Or perhaps you have been unlucky in love and would like to reverse this trend? Do you always seem to be struggling to make ends meet, or feel that you deserve a pay rise? If your answer to any of the above questions is 'yes', then one of these three key rituals could be the answer to your problems!

THE STRESS BUSTER

Because you are focusing on your health and wellbeing, iron pyrites is the perfect talisman for dealing with overall stress and its unpleasant effects.

If your stress is work-related, choose fire opal as your talisman to help you to make good decisions about how to prioritize your workload. Alternatively, try using a rock crystal wand, as this stone might spark a new idea as to how to reorganize your working practices and to communicate your plan to your boss, or someone with the authority to take action.

Prior to your ritual, gather together some yellow and grey-green candles. You could also incorporate these colours into decorations, perhaps using a grey-green tablecloth or backdrop and yellow flowering house plants, for example.

The ritual should be performed on a Sunday and any gold item may be set beside the pyrites. If an incense or perfume with a fresh scent can be used, so much the better.

Iron pyrites is the perfect talisman for dealing with overall stress and its unpleasant effects

Before the ritual, you might like to snack on something sweet. If you have time, you could take the iron pyrites to a public garden or art gallery where it can soak up the atmosphere. Visiting either of these relaxing spots is particularly effective, since you will easily be able to conjure up the image of the place in your mind, and even use it as a refuge when stress is at its worst.

Alternatively, you could place a keepsake, such as a painting, photograph or postcard of a garden or gallery, beside the stone during the rite.

THE WEALTH BRINGER

To attract wealth, luck and opportunity, sodalite is the ideal talisman. However, if you need money particularly urgently, you might be best to use star sapphire or falcon's eye. If you are hoping to soon reap the financial benefits from a long-term project, then sapphire would be your best bet.

Schedule your ritual for a Thursday and choose candles and plan your colour scheme in shades of blue and gold. Gather together any items made of brass and scents or perfumes with a particularly warm, intoxicating odour. Coins and bank notes, being representative of wealth, may also be liberally scattered around.

day for your rite is a Friday, and as green and scarlet are the appropriate colours, the room could be decorated with candles, tablecloths, flowering plants and so on, in these vibrant and beautiful hues.

The scent of citrus would strike the right note and any personal souvenirs of hilltops or sporting competitions would also help to attract and project the right energies.

Any items made of copper will add strength to the rite, which can also be enhanced with personal touches to indicate the kind of lover you are hoping to attract. Red roses, for example, would suggest that you're out to attract a romantic, while a golf ball or tennis racquet would imply that you're keen to catch a sporty partner!

How long before you see results?

As a rule, talismans exert their power continually and at a constant level. They work in a subtle but cumulative way, so although it may take time to see any effect, your talisman will have been working behind the scenes, clearing obstacles and setting up the potential for your goal to be realized.

However, you can also draw off talismanic power from the stones in bursts. When you specifically need their powers, all you have to do is to concentrate on them and reconnect with their special nature.

If, at any time, a stone starts to be intrusive and becomes a distraction, leave it alone for a while. If you are still having problems with it, then the stone should be thoroughly cleansed and re-energized.

Prior to the ritual, take your talisman to a place that symbolizes wealth – a bank, a jewellers, or an expensive hotel, for example. It isn't necessary to go inside if you don't want to, but try to touch the wall of the building. Before performing the ritual, treat your taste buds to some rich food.

THE PATH TO LOVE

The perfect stone for attracting a new lover is peridot. However, if you want to attract the attention of someone you already know, then emerald would be more suitable. The best

Pendulum dowsing

Pendulums offer an inexpensive and absorbing way to reveal and experiment with your psychic abilities. Forked sticks have been used to find underground water for many centuries, and although pendulums may be a recent extension of the ancient tradition of dowsing, they have gained a dedicated and enthusiastic following.

Pendulums can be used for a wide range of purposes, from treasure hunting, prospecting for precious metals, locating missing persons and lost pets, fortune telling, determining the sex of unborn children and fun party games. Part of the appeal of this form of dowsing is the ease with which anybody can investigate and use the power of the pendulum.

MAKING AND HOLDING YOUR PENDULUM

A dedicated dowser might invest in a specially ground and flawless rock crystal pendulum, drilled at one end, and attached to a fine silver chain. However, almost any smallish rock crystal wand will reveal your aptitude for this occult technique. The wand should be about 2–3cm (¾–1in) long and hung by a thread so that the natural facets of the crystal point hang down – pointing to whatever is being dowsed. The thread should be around 30cm (12in) long and may be fixed to the crystal with strong glue. Natural fibres such as cotton, hemp or wool (but not silk) are preferred – white or purple are recommended.

Spectrolite

If your wand has a rough or knobbly base around which the thread (or silver chain) can be secured without using glue, then so much the better. If you plan to use your pendulum out of doors, where strong winds can be a serious handicap, then the pendulum can be heavier and larger (around 6–8cm [2¼–3in] long and hung from a cord or chain approximately half a metre long). In this instance, basal features to which the cord or chain can be tied are particularly helpful.

When holding a pendulum, most people instinctively grip the thread between the forefinger and thumb of their right hand. However, when dowsing for long periods of time or with a heavy pendulum this grip can become tiring, which can interfere with the sensitivity of the instrument.

A popular alternative is to hold the thread in the palm of your left hand (unless you are left-handed, in which case you should use your right hand) using your thumb and all of your fingers except the forefinger.

The forefinger is extended (as if pointing to something) and then slightly crooked. The thread is then simply hooked over the furthermost section of the finger – you should avoid getting it caught in the crease of the joint itself. This arrangement provides maximum sensitivity with a high degree of comfort.

Furthermore, if the pendulum is a little heavy then it may be dangled over the middle section of the forefinger or, if it is particularly large, then you can drape the thread across the strongest bone of the finger, the section closest to the knuckle.

If you are dowsing an object, such as a map, which can be placed on a table then by all means sit down and make yourself comfortable. Many dowsers actually rest their elbow on the table to avoid being distracted by tired arm muscles.

USING YOUR PENDULUM

The pendulum communicates in a very simple way: it either stays still, swings gently to and fro in an arc (which may widen into an oval, and then become a circle), or swing in a full circle right from the start (which may narrow into an oval, and then become an arc). The oval or circle can be described as having a clockwise or anti-clockwise direction.

If you are just beginning to experiment with your abilities, it is worth spending some time just holding the thread and letting the

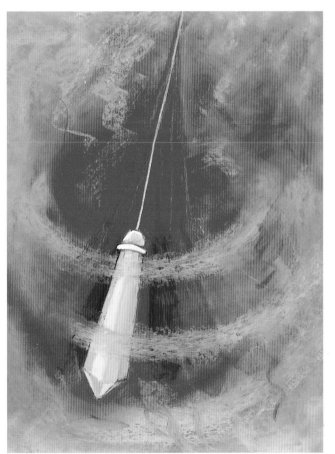

pendulum twirl and gyrate with random motion. This is called the starting position, and gives you time to settle down and feel comfortable with what you are doing. Just watch it for now. Don't try to get it to answer any questions but simply get used to the way it feels to have it dancing freely around.

In most forms of divination there is an accepted way of doing things. Pendulum dowsing is different. Most authorities agree that you must conduct your own experiments to interpret the pendulum's movements: there are no hard and fast rules to follow.

This may sound daunting, but it really is very simple. All you need to do is ask it to respond to a few straightforward questions and, by observing the shape of its answers, you can quickly decipher its language.

It is worth noting that, if you have difficulty in getting the pendulum to move at all, then you can deliberately set it moving in an arc – this will be your starting position – and the answers to your questions will then be determined by the direction (clockwise or anti-clockwise) in which the pendulum starts to move when it begins to form an oval or circle.

GETTING STARTED

The best way to begin is to ask a question to which you would expect to receive a 'yes' response. Hold the pendulum (in the starting position) so that it dangles a couple of centimetres above the palm of your empty hand. If you're a woman, ask the question, 'Am I female?' Or, if you're a man, ask 'Am I male?' Clearly, whatever motion it makes means 'yes'.

Now place a pen on the table and ask the question, 'Is this an apple?' If it seems to make the same motion as before, when it answered 'yes', then clearly something is wrong, and more practise is required. But if it responds in a completely new way then this can be interpreted as the 'no' response.

It is worth noting here that although a few people do seem to have an immediate affinity with dowsing, everybody is advised to check and double-check the responses (using as wide a variety of subjects as possible) before putting their pendulum to use in an important situation.

USING YOUR PENDULUM IN EVERYDAY LIFE

Because the pendulum can amplify minute movements in your arm, hand and fingers, it is a particularly valuable tool for contacting your subconscious. Everyday quandaries such as 'Can I really be bothered to go to the party tonight?' can be put directly to the pendulum: 'Should I go to this party?' The response – 'yes' or 'no' – will come from your subconscious which knows exactly how tired you will be, will have a perfectly clear perception of how good the party is likely to be, and will also be aware of a thousand and one little details that might have slipped your conscious mind, such as 'Oh yes, Abc was invited… and she might bring Xyz, and I'd like to meet Xyz again!' Or you could picture Xyz in your mind and ask: 'Will Xyz be at the party?'

On the subject of parties, why not introduce the pendulum to the gathering – not to demonstrate your own prowess, but to allow other people the opportunity of having a go. You could put an *hors d'oeuvre* under one of three upturned cups, for example, as a reward for a successful dowse! Of course, the range of questions you can put to a pendulum is limited only by the scope of your imagination.

Often your subconscious can warn you against trusting someone who has conned your conscious mind. 'Should I invite Xyz in for a drink next time?' 'Should I put my money into Abc's business scheme?' 'Should I get a doctor to give me a second opinion?' Remember, the final decision is taken by your conscious mind, so there's no need to worry about falling prey to superstitious mumbo jumbo.

In fact, there is more danger of surreptitiously trying to tweak the pendulum thread to give you the response that you consciously hope to see!

It is actually quite important to quieten your emotions before asking any question that raises strong feelings. Experience, perseverance and being honest with yourself will be your best guides in this fascinating voyage of self-exploration.

Using your pendulum on special occasions

A pendulum is commonly used to discern the sex of an unborn child. Whether dowsing directly on the mother-to-be, or a photo of her (or even just using a strongly imagined thought of her), it is prudent to first ask the pendulum: 'Is this adult female?' Once you are satisfied with that initial response you can move your attention to the baby itself: 'Is this baby female?'

Although there is a 50–50 chance of getting the sex right by mere chance, it is an important question, and a wrong diagnosis will not be quickly forgotten by the parents. In such delicate circumstances it is well worthwhile taking pains to double-check the answer by asking, 'Is this baby male?'

Obviously your own subconscious is not in any position to know whether an unborn baby is male or female – so it is a much deeper connection that dowsers are making when they ask questions such as these.

By spreading a map on a table, and slowly moving the pendulum over its whole surface, you can even try to locate missing persons, lost pets, mislaid or even stolen items, buried treasure and valuable mineral deposits, or even healing ley lines, by repeatedly asking: 'Is [name] here?'

Map dowsing is particularly popular as it is so versatile. It can be used to solicit an answer from a map of the world by asking 'Should I spend my holiday here?' Or you can use a sketch of the house and garden to ask 'Is my missing earring here?' or 'Are my lost keys here?'

Furthermore, you can make a series of simple cards, each bearing a single word such as the names of dates, people, places, jobs or anything else that interests you.

Place the cards in a crescent or horseshoe shape, concentrate on the question, and watch towards which card the pendulum swings – that card reveals the answer. You may prefer to shuffle the cards and have them face down, so that you do not consciously influence the pendulum. And do remember to treat these answers as interesting guidance only – not as destiny written in stone!

A FINAL WISH

Crystals and other stones have long held a fascination for the human race. While science explores the physical attributes of crystals (the first laser, for example, had a ruby at its heart), and is poised for greater discoveries to come, we can all undertake a great adventure into the inner mind and explore the world of the spirit – territory to which science has so far been unable to find a gateway. Many people discover that crystals and other stones can be their passport to a great personal adventure.

It is in this spirit of continuous self-discovery that we leave the future in your hands: hold it safely, for who knows what wonders yet may come to be.

*I*ndex of crystals

General index

Acknowledgements

The authors and publishers would like to thank the following for permission to reproduce photographs:

Ardea: p74/75 background, 102 below, 111 above

Bridgeman Art Library: p. 95 above, 98 right, 99, 112, 116

Geoff Dann: p. 39 above, 47 below, 56 below, 60, 63 below, 64 below, 65 below, 66 below

Michelle Garrett: p. 17, 19, 22, 29, 31, 33, 34/35, 37 below, 38 below, 40 above, 41 below, 42, 43 above, 44 above , 45 below, 46, 48, 72

GeoScience Features Library: p. 27, 32 malachite, 77 jet, 115 below

Jon Hamilton: p. 23 imperial topaz, 32, 87 amber, 55, 59 above, 62, 77 tiger's eye, 97, 120

Natural History Museum Picture Library: p. 3, 8/9 main pic, 24/back jacket, 27 sapphire, 32 red jasper, 38 above, 39 below, 56 above, 76/77 sapphire, 119 below, 122 below, 123 below, 127 below, 142 above right

Science Photo Library: p. 6/7, 10/11 background, 12/13 background, 14, 25 background, 27 ruby, 50/51, 52/53 background, 80 (76/77), 100 (76/77), 107 below right, 137 left, 148 below

Jacket photography by Geoff Dann and Michelle Garrett.
All crystal images photographed by Geoff Dann, unless where stated otherwise.

The publishers would like to thank Jonathan Dee for permission to print the Moon Finder chart on page 78.